First Edition, 1957
Second Impression, 1958
Second Edition, 1963
Reprinted 1966

PRINTED IN GREAT BRITAIN
BY A. T. BROOME AND SON, ST. CLEMENT'S, OXFORD

INTENTION

by

G. E. M. ANSCOMBE

Lecturer in Philosophy and
Research Fellow in Somerville College

OXFORD
BASIL BLACKWELL

CONTENTS

INTRODUCTION

The greater part of what appears here was delivered as a course of lectures at Oxford in the Hilary Term of 1957. Excerpts, with small modifications, comprising the discussion of the difference between ' motive ', ' intention ' and ' mental cause ' formed an Aristotelian Society paper delivered on June 3rd, 1957. I am indebted to the Society for permission for a substantial reprint of that matter. This book assembles the results, so far as concerns this particular topic, of research begun during my tenure of the Mary Somerville Research Fellowship at Somerville College. I wish therefore to express my gratitude to the Donors. More recently I have been supported by the Rockefeller Foundation, to which an acknowledgment is therefore also due.

Note on the Second Impression

I have made a few alterations; the only ones of any significance are on pp. 29, 58, 59 and 61.

Note on Second Edition

For this edition I have made some small alterations in §§ 2, 6, 17, 33 and 34.

INTENTION

1.   Very often, when a man says ' I am going to do such-and-such ', we should say that this was an expression of intention. We also sometimes speak of an action as intentional, and we may also ask with what intention the thing was done. In each case we employ a concept of ' intention '; now if we set out to describe this concept, and took only one of these three kinds of statement as containing our whole topic, we might very likely say things about what ' intention ' means which it would be false to say in one of the other cases. For example, we might say ' Intention always concerns the future '. But an action can be intentional without being concerned with the future in any way. Realising this might lead us to say that there are various senses of ' intention ', and perhaps that it is thoroughly misleading that the word ' intentional ' should be connected with the word ' intention ', for an action can be intentional without having any intention in it. Or alternatively we may be tempted to think that only actions done with certain further intentions ought to be called intentional. And we may be inclined to say that ' intention ' has a different sense when we speak of a man's intentions *simpliciter*— i.e. what he intends to do—and of his intention *in* doing or proposing something—what he aims at in it. But in fact it is implausible to say that the word is equivocal as it occurs in these different cases.

Where we are tempted to speak of ' different senses ' of a word which is clearly not equivocal, we may infer that we are in fact pretty much in the dark about the character of the concept which it represents. There is, however, nothing wrong with taking a topic piecemeal. I shall therefore begin my enquiry by considering expressions of intention.

2.   The distinction between an expression of intention and a prediction is generally appealed to as something intuitively clear. ' I am going to be sick ' is usually a prediction; ' I am going to take a walk ' usually an expression of intention. The distinction intended *is* intuitively clear, in the following sense: if

I say ' I am going to fail in this exam. ' and someone says ' Surely you aren't as bad at the subject as that ', I may make my meaning clear by explaining that I was expressing an intention, not giving an estimate of my chances.

If, however, we ask in philosophy what the difference is between e.g. ' I am going to be sick ' as it would most usually be said, and ' I am going to take a walk ', as *it* would most usually be said, it is not illuminating to be told that one is a prediction and the other the expression of an intention. For we are really asking what each of these is. Suppose it is said 'A prediction is a statement about the future '. This suggests that an expression of intention is not. It is perhaps the description—or expression—of a present state of mind, a state which has the properties that characterise it as an intention. Presumably what these are has yet to be discovered. But then it becomes difficult to see why they should be essentially connected with the future, as the intention seems to be. No one is likely to believe that it is an accident, a mere fact of psychology, that those states of mind which are intentions always have to do with the future, in the way that it is a fact of racial psychology, as one might say, that most of the earliest historical traditions concern heroic figures. And if you try to make being concerned with the fut ure into a defining property of intentions, you can be asked what serves to distinguish this concern with the future from the predictive concern.

Let us then try to give some account of prediction. The following seems promising: a man says something with one inflection of the verb in his sentence; later that same thing, only with a changed inflection of the verb, can be called true (or false) in face of what has happened later.

Now by this criterion, commands and expressions of intention will also be predictions. In view of the difficulties described above, this may not constitute an objection. Adopting a hint from Wittgenstein (*Philosophical Investigations* §§ 629–30) we might then first define prediction in general in some such fashion, and then, among predictions, distinguish between commands, expressions of intention, estimates, pure prophecies, etc. The ' intuitively clear ' distinction we spoke of turns out to be a distinction between expressions of intention and estimates. But

I say ' I am going to fail in this exam. ' and someone says ' Surely you aren't as bad at the subject as that ', I may make my meaning clear by explaining that I was expressing an intention, not giving an estimate of my chances.

If, however, we ask in philosophy what the difference is between e.g. ' I am going to be sick ' as it would most usually be said, and ' I am going to take a walk ', as *it* would most usually be said, it is not illuminating to be told that one is a prediction and the other the expression of an intention. For we are really asking what each of these is. Suppose it is said 'A prediction is a statement about the future '. This suggests that an expression of intention is not. It is perhaps the description—or expression—of a present state of mind, a state which has the properties that characterise it as an intention. Presumably what these are has yet to be discovered. But then it becomes difficult to see why they should be essentially connected with the future, as the intention seems to be. No one is likely to believe that it is an accident, a mere fact of psychology, that those states of mind which are intentions always have to do with the future, in the way that it is a fact of racial psychology, as one might say, that most of the earliest historical traditions concern heroic figures. And if you try to make being concerned with the future into a defining property of intentions, you can be asked what serves to distinguish this concern with the future from the predictive concern.

Let us then try to give some account of prediction. The following seems promising: a man says something with one inflection of the verb in his sentence; later that same thing, only with a changed inflection of the verb, can be called true (or false) in face of what has happened later.

Now by this criterion, commands and expressions of intention will also be predictions. In view of the difficulties described above, this may not constitute an objection. Adopting a hint from Wittgenstein (*Philosophical Investigations* §§ 629–30) we might then first define prediction in general in some such fashion, and then, among predictions, distinguish between commands, expressions of intention, estimates, pure prophecies, etc. The ' intuitively clear ' distinction we spoke of turns out to be a distinction between expressions of intention and estimates. But

INTRODUCTION

The greater part of what appears here was delivered as a course of lectures at Oxford in the Hilary Term of 1957. Excerpts, with small modifications, comprising the discussion of the difference between ' motive ', ' intention ' and ' mental cause ' formed an Aristotelian Society paper delivered on June 3rd, 1957. I am indebted to the Society for permission for a substantial reprint of that matter. This book assembles the results, so far as concerns this particular topic, of research begun during my tenure of the Mary Somerville Research Fellowship at Somerville College. I wish therefore to express my gratitude to the Donors. More recently I have been supported by the Rockefeller Foundation, to which an acknowledgment is therefore also due.

Note on the Second Impression

I have made a few alterations; the only ones of any significance are on pp. 29, 58, 59 and 61.

Note on Second Edition

For this edition I have made some small alterations in §§ 2, 6, 17, 33 and 34.

a single utterance may function as more than one of these kinds
of prediction. E.g. when a doctor says to a patient in the presence
of a nurse 'Nurse will take you to the operating theatre', this
may function both as an expression of his intention (if it is in
it that his decision as to what shall happen gets expressed) and
as an order, as well as being information to the patient; and it is
this latter in spite of being in no sense an estimate of the future
founded on evidence, nor yet a guess or prophecy; nor does the
patient normally *infer* the information from the fact that the
doctor said that; he would say that the doctor *told* him. This
example shews that the indicative (descriptive, informatory)
character is not the distinctive mark of 'predictions' *as opposed
to* 'expressions of intention', as we might at first sight have been
tempted to think.

An imperative will be a description of some future action,
addressed to the prospective agent, and cast in a form whose
point in the language is to make the person do what is described.
I say that this is its point in the language, rather than that it is
the purpose of the speaker, partly because the speaker might of
course give an order with some purpose quite other than that
it should be executed (e.g. so that it should *not* be executed),
without detriment to its being an order.

Execution-conditions for commands correspond to truth-
conditions for propositions. What are the reasons other than
a dispensable usage for not calling commands true and false
according as they are obeyed or disobeyed?

An order will usually be given with some intention or other,
but is not as such the expression of a volition; it is simply a
description of an action cast in a special form; this form is
sometimes a special inflection and sometimes a future tense
which has other uses as well.

Orders are usually criticised for being sound or unsound
rather than for being fulfilled or not fulfilled; but this does not
serve to distinguish orders from estimates of the future, since the
same may hold for estimates of the future, where these are
scientific. (Unscientific estimates are of course praised for being
fulfilled rather than for being well-founded, as no one knows
what a good foundation is for an unscientific estimate—e.g. a
political one.) But there is a difference between the types of

ground on which we call an order, and an estimate of the future, sound. The reasons justifying an order are not ones suggesting what is probable, or likely to happen, but e.g. ones suggesting what it would be good to make happen with a view to an objective, or with a view to a sound objective. In this regard, commands and expressions of intention are similar.

It is natural to feel an objection both to calling commands, and to calling expressions of intention, predictions. In the case of commands, the reason lies in the superficial grammar, and just because of this is more easily disposed of. In the case of intentions, superficial grammar would rather incline us to accept the diagnosis, since a common form of expression of intention is a simple future tense, and indeed, this use of the future tense must play a dominant part in any child's learning of it. But our objections are deeper rooted.

If I do not do what I said I would, I am not supposed to have made a mistake, or even necessarily to have lied; so it seems that the truth of a statement of intention is not a matter of my doing what I said. But why should we not say: this only shows that there are other ways of saying what is not true, besides lying and being mistaken?

A lie, however, is possible here; and if I lie, what I say is a lie because of something present, not future. I might even be lying in saying I was going to do something, though I afterwards did it. The answer to this is that a lie is an utterance contrary to one's mind, and one's mind may be either an opinion, or a mind to make something the case. That a lie is an utterance contrary to one's mind does not mean that it is a false report of the contents of one's mind, as when one lies in response to the query 'A penny for your thoughts'.

One might not have a ' mind ' to do something, distinguishable from uttering the words. And then, as Quine once put it (at a philosophical meeting), one might do the thing 'to make an honest proposition ' of what one had said. For if I don't do what I said, what I said was not true (though there might not be a question of my *truthfulness* in saying it). But the reason why Quine's remark is a joke is that this falsehood does not necessarily impugn what I *said*. In some cases the facts are, so to speak, impugned for not being in accordance with the words, rather than *vice versa*. This is sometimes so when I change my mind; but another case of it occurs when e.g. I write something other than I think I am writing: as Theophrastus says (*Magna Moralia*,[1] 1189b 22), the mistake here is one of performance, not of judgment. There are other cases too: for example, St. Peter did not *change his mind* about denying Christ; and yet it would not be correct to say he made a lying promise of faithfulness.

A command is essentially a sign (or symbol), whereas an intention can exist without a symbol; hence we speak of commands, not of the expression of commanding; but of the *expression of* intention. This is another reason for the very natural idea that in order to understand the expression of intention, we ought to consider something internal, i.e. what it is an expression of. This consideration disinclines us to call it a prediction—i.e. a description of something future. Even though that is just what ' I'll do such-and-such ' actually looks like, and even though ' I intend to go for a walk but shall not go for a walk ' does sound in some way contradictory.

Intention appears to be something that we can express, but which brutes (which e.g. do not give orders) can *have*, though lacking any distinct expression of intention. For a cat's movements in stalking a bird are hardly to be called an expression of intention. One might as well call a car's stalling the *expression* of its being about to stop. Intention is unlike emotion in this respect, that the expression of it is purely conventional; we might say ' linguistic ', if we will allow certain bodily movements with a conventional meaning to be included in language. Wittgenstein seems to me to have gone wrong in speaking of the ' natural expression of an intention ' (*Philosophical Investigations* § 647).

3. We need a more fruitful line of enquiry than that of considering the verbal expression of intention, or of trying to consider what it is an expression of. For if we consider just the verbal expression of intention, we arrive only at its being a— queer—species of prediction; and if we try to look for what it is an expression of, we are likely to find ourselves in one or other of several dead ends, e.g.: psychological jargon about ' drives ' and

---

[1] Assuming that we are correctly told that Theophrastus was the author.

' sets '; reduction of intention to a species of desire, i.e. a kind of emotion; or irreducible intuition of the meaning of ' I intend '.

Looking at the verbal expression of intention is indeed of use for avoiding these particular dead-ends. They are all reached in consequence of leaving the distinction between estimation of the future and expression of intention as something that just is intuitively obvious. A man says ' I am going for a walk ' and we say ' that is an expression of intention, not a prediction '. But how do we know? If we asked him, no doubt he would tell us; but what does he know, and how? Wittgenstein has shown the impossibility of answering this question by saying ' He recognizes himself as having, or as having had, an intention of going for a walk, or as having meant the words as an expression of intention '. If this were correct, there would have to be room for the possibility that he misrecognizes. Further, when we remember having meant to do something, what memory reveals as having gone on in our consciousness is a few scanty items at most, which by no means add up to such an intention; or it simply prompts us to use the words ' I meant to . . . ', without even a mental picture of which we judge the words to be an appropriate description. The distinction, then, cannot be left to be intuitively obvious, except where it is used to answer the question in what sense a man meant the form of words ' I am going to . . . ' on a particular occasion.

We might attempt to make the distinction out by saying: an expression of intention is a description of something future in which the speaker is some sort of agent, which description he justifies (if he does justify it) by reasons for acting, sc. reasons why it would be useful or attractive if the description came true, not by evidence that it is true. But having got so far, I can see nowhere else to go along this line, and the topic remains rather mystifying. I once saw some notes on a lecture of Wittgenstein in which he imagined some leaves blown about by the wind and saying ' Now I'll go this way . . . now I'll go that way ' as the wind blew them. The analogy is unsatisfactory in apparently assigning no role to these predictions other than that of an unnecessary accompaniment to the movements of the leaves. But it might be replied: what do you mean by an ' unnecessary ' accompaniment? If you mean one in the absence of which the movements of the leaves would have been just the same, the analogy is certainly bad. But how do you know what the movements of the leaves would have been if they had not been accompanied by those thoughts? If you mean that you could calculate their movements just by knowing the speed and direction of the winds and the weight and other properties of the leaves, are you insisting that such calculations could not include calculations of their thoughts?—Wittgenstein was discussing free will when he produced this analogy; now the objection to it is not that it assigns a false role to our intentions, but only that it does not describe their role at all; this, however, was not its purpose. That purpose was clearly *some* denial of free will, whether we take the wind as a symbol for the physical forces that affect us, or for God or fate. Now it may be that a correct description of the role of intention in our actions will not be relevant to the question of free will; in any case I suspect that this was Wittgenstein's view; therefore in giving this anti-freewill picture he was at liberty simply to leave the role of intention quite obscure.

Now our account of expressions of intention, whereby they are distinguished from estimates of the future, leaves one in very much the same position as does the picture of the wind blowing the leaves. People do in fact give accounts of future events in which they are some sort of agents; they do not justify these accounts by producing reasons why they should be believed but, if at all, by a different sort of reason; and these accounts are very often correct. This sort of account is called an expression of intention. It just does occur in human language. If the concept of ' intention ' is one's quarry, this enquiry has produced results which are indeed not false but rather mystifying. What is meant by ' reason ' here is obviously a fruitful line of enquiry; but I prefer to consider this first in connexion with the notion of intentional action.

4.   I therefore turn to a new line of enquiry: how do we tell someone's intentions? or: what kind of true statements about people's intentions can we certainly make, and how do we know that they are true? That is to say, is it possible to find types of statement of the form 'A intends X ' which we can say have a

great deal of certainty? Well, if you want to say at least some true things about a man's intentions, you will have a strong chance of success if you mention what he actually did or is doing. For whatever else he may intend, or whatever may be his intentions in doing what he does, the greater number of the things which you would say straight off a man did or was doing, will be things he intends.

I am referring to the sort of things you would say in a law court is you were a witness and were asked what a man was doing when you saw him. That is to say, in a very large number of cases, your selection from the immense variety of true statements about him which you might make would coincide with what he could say he was doing, perhaps even without reflection, certainly without adverting to observation. I am sitting in a chair writing, and anyone grown to the age of reason in the same world would know this as soon as he saw me, and in general it would be his first account of what I was doing; if this were something he arrived at with difficulty, and what he knew straight off were precisely how I was affecting the acoustic properties of the room (to me a very recondite piece of information), then communication between us would be rather severely impaired.

In this way, with a view to shewing roughly the range of things to be discovered here, I can take a short cut here, and discuss neither how I am to select from the large number of true statements I could make about a person, nor what is involved in the existence of such a straight-off description as ' She is sitting in a chair and writing '. (Not that this does not raise very interesting questions. See *Philosophical Investigations*, p. 59, (*b*): ' I see a picture: it shows a man leaning on a stick and going up a steep path. How come? Couldn't it look like that if he were sliding downhill in that position? Perhaps a Martian would give that description.' *Et passim*.) All I am here concerned to do is note the fact: we can simply say 'Look at a man and say what he is doing'—i.e. say what would immediately come to your mind as a report to give someone who could not see him and who wanted to know what was to be seen in that place. In most cases what you will say is that the man himself knows; and again in most, though indeed in fewer, cases you will be reporting not merely what he is doing, but *an* intention of his—namely, to do that thing. What is more, if it is not an intention of his, this will for the most part be clear without asking him.

Now it can easily seem that in general the question what a man's intentions are is only authoritatively settled by him. One reason for this is that in general we are interested, not just in a man's intention *of* doing what he does, but in his intention *in* doing it, and this can very often not be seen from seeing what he does. Another is that in general the question whether he intends to do what he does just does not arise (because the answer is obvious); while if it does arise, it is rather often settled by asking him. And, finally, a man can form an intention which he then does nothing to carry out, either because he is prevented or because he changes his mind: but the intention itself can be complete, although it remains a purely interior thing. All this conspires to make us think that if we want to know a man's intentions it is into the contents of his mind, and only into these, that we must enquire; and hence, that if we wish to understand what intention is, we must be investigating something whose existence is purely in the sphere of the mind; and that although intention issues in actions, and the way this happens also presents interesting questions, still what physically takes place, i.e. what a man actually does, is the very last thing we need consider in our enquiry. Whereas I wish to say that it is the first. With this preamble to go on to the second head of the division that I made in § 1 : intentional action.

5. What distinguishes actions which are intentional from those which are not? The answer that I shall suggest is that they are the actions to which a certain sense of the question ' Why? ' is given application; the sense is of course that in which the answer, if positive, gives a reason for acting. But this is not a sufficient statement, because the question " What is the relevant sense of the question ' Why? ' " and " What is meant by ' reason for acting ' ? " are one and the same.

To see the difficulties here, consider the question, ' Why did you knock the cup off the table? ' answered by ' I thought I saw a face at the window and it made me jump '. Now, so far I have only characterised reason for acting by opposing it to evidence for supposing the thing will take place—but the ' reason ' here was not evidence that I was going to knock the cup off the table. Nor can we say that since it mentions something previous to the action, this will be a cause rather than a reason; for if you ask ' Why did you kill him? ' the answer ' He killed my father ' is surely a reason rather than a cause, but what it mentions is previous to the action. It is true that we don't ordinarily think of a case like giving a sudden start when we speak of a reason for acting. " Giving a sudden start ", someone might say, " is not *acting* in the sense suggested by the expression ' reason for acting '. Hence, though indeed we readily say e.g. ' What was the reason for your starting so violently? ' this is totally unlike ' What is your reason for excluding so-and-so from your will? ' or ' What is your reason for sending for a taxi? '" But what *is* the difference? In neither case is the answer a piece of evidence. Why is giving a start or gasp not an ' action ', while sending for a taxi, or crossing the road, is one? The answer cannot be " Because the answer to the question ' why? ' may give a *reason* in the latter cases ", for the answer may ' give a reason ' in the former cases too; and we cannot say "Ah, but not a reason for *acting* "; we should be going round in circles. We need to find the difference between the two kinds of ' reason ' without talking about ' acting '; and if we do, perhaps we shall discover what is meant by ' acting ' when it is said with this special emphasis.

It will hardly be enlightening to say : in the case of the sudden start the ' reason ' is a *cause*; the topic of causality is in a state of too great confusion; all we know is that this is one of the places where we do use the word ' cause'. But we also know that this is a rather strange case of causality; the subject is able to give the cause of a thought or feeling or bodily movement in the same kind of way as he is able to state the place of his pain or the position of his limbs.

Nor can we say: "—Well, the ' reason ' for a movement is a cause, and not a reason in the sense of ' reason for acting ', when the movement is involuntary; it is a reason, as opposed to a cause, when the movement is voluntary and intentional.' This is partly because in any case the object of the whole enquiry is really to delineate such concepts as the voluntary and the intentional, and partly because one can also give a ' reason ' which is only a ' cause ' for what is voluntary and intentional. E.g. " Why are you walking up and down like that?" —"It's that military band; it excites me". Or "What made you sign the document at last?"—"The thought: 'It is my duty' kept hammering away in my mind until I said to myself 'I can do no other', and so signed."

It is very usual to hear that such-and-such are what we *call* 'reasons for acting' and that it is 'rational' or 'what we *call* rational' to act for reasons; but these remarks are usually more than half moralistic in meaning (and moralism, as Bradley remarked, is bad for thinking); and for the rest they leave our conceptual problems untouched, while pretending to give a quick account. In any case, this pretence is not even plausible, since such remarks contain no hint of what it is to act for reasons.

6. To clarify the proposed account, "Intentional actions are ones to which a certain sense of the question 'why?' has application", I will both explain this sense and describe cases shewing the question *not* to have application. I will do the second job in two stages because what I say in the first stage of it will be of use in helping to explain the relevant sense of the question 'why?'.

This question is refused application by the answer: 'I was not aware I was doing that'. Such an answer is, not indeed a proof (since it may be a lie), but a claim, that the question 'Why did you do it (are you doing it)?', in the required sense, has no application. It cannot be plausibly given in every case; for example, if you saw a man sawing a plank and asked 'Why are you sawing that plank?', and he replied 'I didn't know I was sawing a plank', you would have to cast about for what he might mean. Possibly he did not know the word 'plank' before, and chooses this way of expressing that. But this question as to what he might mean need not arise at all—e.g. if you ask someone why he is standing on a hose-pipe and he says 'I didn't know I was'.

Since a single action can have many different descriptions, e.g. 'sawing a plank', 'sawing oak', 'sawing one of Smith's planks', 'making a squeaky noise with the saw', 'making a great deal of sawdust' and so on and so on, it is important to notice that a man may know that he is doing a thing under one description, and not under another. Not every case of this is a case of his knowing that he is doing one part of what he is doing and not another (e.g. he knows he is sawing but not that he is making a squeaky noise with the saw). He may know that he is sawing a plank, but not that he is sawing an oak plank or Smith's plank; but sawing an oak plank or Smith's plank is not something else that he is doing besides just sawing the plank that he is sawing. For this reason, the statement that a man knows he is doing X does not imply the statement that, concerning anything which is also his doing X, he knows that he is doing that thing. So to say that a man knows he is doing X is to give a description of what he is doing *under which* he knows it. Thus, when a man says ' I was not aware that I was doing X ', and so claims that the question ' Why? ' has no application, he cannot always be confuted by the fact that he was attentive to those of his own proceedings in which doing X consisted.

7. It is also clear that one is refusing application to the question ' Why? ' (in the relevant sense) if one says: ' It was involuntary ', even though the action was something of which one was aware. But I cannot use this as it stands, since the notion of the involuntary pretty obviously covers notions of exactly the type that a philosophical enquiry into intention ought to be elucidating.

Here, digressing for a moment, I should like to reject a fashionable view of the terms ' voluntary ' and ' involuntary ', which says they are appropriately used only when a person has done something untoward. If anyone is tempted by this view, he should consider that physiologists are interested in voluntary action, and that they are not giving a special technical sense to the word. If you ask them what their criterion is, they say that if they are dealing with a grown human they ask him, and if with an animal, they take movements in which the animal is e.g. trying to get at something, say food. That is, the movement by which a dog cocked its ear at a sudden sound would not be used as an example.

This does not mean that every description of action in which its voluntariness can be considered is of interest to physiologists. Of course they are only interested in bodily movements.

We can also easily get confused by the fact that ' involuntary ' neither means simply non-voluntary, nor has an unproblematic sense of its own. In fact this pair of concepts is altogether very confusing. Consider the four following examples of the involuntary:

(*a*) The peristaltic movement of the gut.

(*b*) The odd sort of jerk or jump that one's whole body sometimes gives when one is falling asleep.

(*c*) ' He withdrew his hand in a movement of involuntary recoil.'

(*d*) ' The involuntary benefit I did him by a stroke I meant to harm him.'

Faced with examples like (*c*) and (*d*), how can I introduce ' It was involuntary ' as a form for rejecting the question ' Why? ' in the special sense which I want to elucidate—when the whole purpose of the elucidation is to give an account of the concept ' intentional '? Obviously I cannot. There is however a class of the things that fall under the concept ' involuntary ', which it is possible to introduce without begging any questions or assuming that we understand notions of the very type I am professing to investigate. Example (*b*) belongs to this class, which is a class of bodily movements in a purely physical description. Other examples are tics, reflex kicks from the knee, the lift of the arm from one's side after one has leaned heavily with it up against a wall.

8. What is required is to describe this class without using any notions like ' intended ' or ' willed ' or ' voluntary ' and ' involuntary '. This can be done as follows: we first point out a particular class of things which are true of a man: namely the class of things which he *knows without observation*. E.g. a man usually knows the position of his limbs without observation. It is without observation, because nothing *shews* him the position of his limbs; it is not as if he were going by a tingle in his knee, which is the sign that it is bent and not straight. Where we can speak of separately describable sensations, having which is in some sense our criterion for saying something, then we can speak of observing that thing; but that is not generally so when we know the position of our limbs. Yet, without prompting, we *can say* it. I say however that we *know* it and not merely *can say*

it, because there is a possibility of being right or wrong: there is point in speaking of knowledge only where a contrast exists between ' he *knows* ' and ' he (merely) *thinks* he knows '. Thus, although there is a similarity between giving the position of one's limbs and giving the place of one's pain, I should wish to say that one ordinarily *knows* the position of one's limbs, without observation, but not that being able to say where one feels pain is a case of something known. This is not because the place of pain (the feeling, not the damage) has to be accepted by someone I tell it to; for we can imagine circumstances in which it is not accepted. As e.g. if you say that your foot, not your hand, is very sore, but it is your hand you nurse, and you have no fear of or objection to an inconsiderate handling of your foot, and yet you point to your foot as the sore part: and so on. But here we should say that it was difficult to guess what you could mean. Whereas if someone says that his leg is bent when it is straight, this may be surprising but is not particularly obscure. He is wrong in what he says, but not unintelligible. So I call this sort of being able to say ' knowledge ' and not *merely* ' being able to say '.

Now the class of things known without observation is of general interest to our enquiry because the class of intentional actions is a sub-class of it. I have already said that ' I was not aware I was doing that ' is a rejection of the question ' Why? ' whose sense we are trying to get at; here I can further say ' I knew I was doing that, but only because I observed it ' would also be a rejection of it. E.g. if one noticed that one operated the traffic lights in crossing a road.

But the class of things known without observation is also of special interest in this part of our enquiry, because it makes it possible to describe the particular class of ' involuntary actions ' which I have so far indicated just by giving a few examples: these are actions like the example (*b*) above, and our task is to mark off this class without begging the questions we are trying to answer. Bodily movements like the peristaltic movement of the gut are involuntary; but these do not interest us, for a man does not know his body is making them except by observation, inference, etc. The involuntary that interests us is restricted to the class of things known without observation; as you would know even with your eyes shut that you had kicked when the doctor tapped your knee, but cannot identify a sensation by which you know it. If you speak of ' that sensation which one has in reflex kicking, when one's knee is tapped ', this is not like e.g. ' the sensation of going down in a lift '. For though one might say ' I thought I had given a reflex kick, when I hadn't moved ' one would never say e.g. ' Being told startling news gives one that sensation': the sensation is not separable, as the sensation ' like going down in a lift ' is.

Now among things known without observation must be included the causes of some movements. E.g. ' Why did you jump back suddenly like that? ' ' The leap and loud bark of that crocodile made me jump '. (I am not saying I did not observe the crocodile barking; but I did not observe that making me jump.) But in examples like (b) the cause of motion is known *only* through observation.

This class of involuntary actions, then, is the class of movements of the body, in a purely physical description, which are known without observation, and where there is no such thing as a *cause* known without observation. (Thus my jump backwards at the leap and bark of the crocodile does *not* belong to this subclass of involuntary actions.) This subclass can be described without our first having clarified the concept ' involuntary '. To assign a movement to it will be to reject the question ' Why? '

9. I first, in considering expressions of intention, said that they were predictions justified, if at all, by a reason for acting, as opposed to a reason for thinking them true. So I here already distinguished a sense of ' Why? ', in which the answer mentions evidence. ' There will be an eclipse tomorrow '.—' Why? ' ' Because . . . '—and an answer is the reason for thinking so. Or ' There was an ancient British camp here '. ' Why?'—and an answer is the reason for thinking so. But as we have already noted, an answer to the question ' Why ?' which does not give reason for thinking the thing true does not *therefore* give a reason for acting. It may mention a cause, and this is far from what we want. However we noticed that there are contexts in which there is some difficulty in describing the distinction between a

cause and a reason. As e.g. when we give a ready answer to the question ' Why did you knock the cup off the table? '—' I saw such-and-such and *it made me jump.* '

Now we can see that the cases where this difficulty arises are just those where the cause itself *qua* cause (or perhaps one should rather say: the causation itself) is in the class of things known without observation.

10.   I will call the type of cause in question a '*mental* cause '. Mental causes are possible, not only for actions (' The martial music excites me, that is why I walk up and down') but also for feelings and even thoughts. In considering actions, it is important to distinguish between mental causes and motives; in considering feelings, such as fear or anger, it is important to distinguish between mental causes and objects of feeling. To see this, consider the following cases:

A child saw a bit of red stuff on a turn in a stairway and asked what it was. He thought his nurse told him it was a bit of Satan and felt dreadful fear of it. (No doubt she said it was a bit of satin.) What he was frightened of was the bit of stuff; the cause of his fright was his nurse's remark. The object of fear may be the cause of fear, but, as Wittgenstein[1] remarks, is not *as such* the cause of fear. (A hideous face appearing at the window would of course be both cause and object, and hence the two are easily confused). Or again, you may be angry *at* someone's action, when what *makes* you angry is some reminder of it, or someone's telling you of it.

This sort of cause of a feeling or reaction may be reported by the person himself, as well as recognised by someone else, even when it is not the same as the object. Note that this sort of causality or sense of ' causality ' is so far from accommodating itself to Hume's explanations that people who believe that Hume pretty well dealt with the topic of causality would entirely leave it out of their calculations; if their attention were drawn to it they might insist that the word ' cause ' was inappropriate or was quite equivocal. Or conceivably they might try to give a Humian account of the matter as far as concerned the outside observer's recognition of the cause; but hardly for the patient's.

[1] *Philosophical Investigations* § 476.

11.   Now one might think that when the question 'Why?' is answered by giving the intention with which a person acts —for example by mentioning something future—this is also a case of a mental cause. For couldn't it be recast in the form: ' Because I wanted . . . ' or ' Out of a desire that . . . '? If a feeling of desire to eat apples affects me and I get up and go to a cupboard where I think there are some, I might answer the question what led to this action by mentioning the desire as having made me . . . etc. But it is not in all cases that ' I did so and so in order to . . . ' can be backed up by ' I *felt* a desire that . . . '. I may e.g. simply hear a knock on the door and go downstairs to open it without experiencing any such desire. Or suppose I feel an upsurge of spite against someone and destroy a message he has received so that he shall miss an appointment. If I describe this by saying ' I wanted to make him miss that appointment ', this does not necessarily mean that I had the thought ' If I do this, he will . . . ' and that affected me with a desire of bringing it about, which led up to my doing so. This may have happened, but need not. It could be that all that happened was this: I read the message, had the thought ' That unspeakable man! ' with feelings of hatred, tore the message up, and laughed. Then if the question ' Why did you do that? ' is put by someone who makes it clear that he wants me to mention the mental causes— e.g. what went on in my mind and issued in the action—I should perhaps give this account; but normally the reply would be no such thing. That particular enquiry is not very often made. Nor do I wish to say that it always has an answer in cases where it can be made. One might shrug or say ' I don't know that there was any definite history of the kind you mean ', or ' It merely occurred to me. . . .'

A ' mental cause ', of course, need not be a mental event, i.e. a thought or feeling or image; it might be a knock on the door. But if it is not a mental event, it must be something perceived by the person affected—e.g. the knock on the door must be heard —so if in this sense anyone wishes to say it is always a mental event, I have no objection. A mental cause is what someone would describe if he were asked the specific question: what produced this action or thought or feeling on your part: what did you see or hear or feel, or what ideas or images cropped up in

C

your mind, and led up to it? I have isolated this notion of a mental cause because there is such a thing as this question with this sort of answer, and because I want to distinguish it from the ordinary senses of ' motive ' and ' intention ', rather than because it is in itself of very great importance; for I believe that it is of very little. But it is important to have a clear idea of it, partly because *a* very natural conception of ' motive ' is that it is what *moves* (the very word suggests that)—glossed as ' what *causes* ' a man's actions etc. And ' what causes ' them is perhaps then thought of as an event that brings the effect about—though how it does—i.e. whether it should be thought of as a kind of pushing in another medium, or in some other way—is of course completely obscure.

12. In philosophy a distinction has sometimes been drawn between our motives and our intentions in acting as if they were quite different things. A man's intention is *what* he aims at or chooses; his motive is what determines the aim or choice; and I suppose that ' determines ' must here be another word for ' causes '.

Popularly motive and intention are not treated as so distinct in meaning. E.g. we hear of ' the motive of gain '; some philosophers have wanted to say that such an expression must be elliptical; gain must be the *intention*, and *desire of gain* the motive. Asked for a motive, a man might say ' I wanted to . . . ', which would please such philosophers; or ' I did it in order to . . . ', which would not; and yet the meaning of the two phrases is here identical. When a man's motives are called good, this may be in no way distinct from calling his intentions good—e.g. he only wanted to make peace among his relations.

Nevertheless there is even popularly a distinction between the meaning of ' motive ' and the meaning of ' intention '. E.g. if a man kills someone, he may be said to have done it out of love and pity, or to have done it out of hatred; these might indeed be cast in the forms ' to release him from this awful suffering ', or ' to get rid of the swine '; but though these are forms of expression suggesting objectives, they are perhaps expressive of the spirit in which the man killed rather than descriptive of the end to which the killing was a means—a future state of affairs to be produced by the killing. And this shows us part of the distinction that there is between the popular senses of motive and intention. We should say: popularly, ' motive for an action ' has a rather wider and more diverse application than ' intention with which the action was done '.

When a man says what his motive was, speaking popularly, and in a sense in which ' motive ' is not interchangeable with ' intention ', he is not giving a ' mental cause ' in the sense that I have given to that phrase.—The fact that the mental causes were such-and-such may indeed help to make his claim intelligible. And further, though he may say that his motive was this or that one straight off and without lying—i.e. without saying what he knows or even half knows to be untrue—yet a consideration of various things, which may include the mental causes, might possibly lead both him and other people to judge that his declaration of his own motive was false. But it appears to me that the mental causes are seldom more than a very trivial item among the things that it would be reasonable to consider. As for the importance of considering the motives of an action, as opposed to considering the intention, I am very glad not to be writing either ethics or literary criticism, to which this question belongs.

Motives may explain actions to us; but that is not to say that they ' determine ', in the sense of causing, actions. We do say: ' His love of truth caused him to . . . ' and similar things, and no doubt such expressions help us to think that a motive must be what produces or brings about a choice. But this means rather ' He did this in that he loved the truth '; it interprets his action.

Someone who sees the confusions involved in radically distinguishing between motives and intentions and in defining motives, so distinct, as the determinants of choice, may easily be inclined to deny both that there is any such thing as mental causality, and that ' motive ' means anything but intention. But both of these inclinations are mistaken. We shall create confusion if we do not notice (a) that phenomena deserving the name of mental causality exist, for we can make the question ' Why? ' into a request for the sort of answer that I considered under that head; (b) that mental causality is not restricted to choices or voluntary or intentional actions, but is of wider application; it is restricted to the wider field of things the agent knows about *not* as an observer, so that it includes some involuntary actions; (c) that motives are not mental causes; and (d) that there is an application for ' motive ' other than the applications of ' the intention with which a man acts '.

13.   Revenge and gratitude are motives; if I kill a man as an act of revenge I may say I do it in order to be revenged, or that revenge is my object; but revenge is not some further thing obtained by killing him, it is rather that killing him is revenge. Asked why I kill him, I reply ' Because he killed my brother '. We might compare this answer, which describes a concrete past event, to the answer describing a concrete future state of affairs which we sometimes get in statements of objectives. It is the same with gratitude, and remorse, and pity for something specific. These motives differ from, say, love or curiosity or despair in just this way: something that *has happened* (or is at present happening) is given as the ground of an action or abstention that is good or bad for the person (it may be oneself, as with remorse) at whom it is aimed. And if we wanted to explain e.g. revenge, we should say it was harming someone because he had done one some harm; we should not need to add to this a description of the feelings prompting the action or of the thought that had gone with it. Whereas saying that someone does something out of, say, friendship cannot be explained in any such way. I will call revenge and gratitude and remorse and pity backward-looking motives, and contrast them with motive-in-general.

Motive-in-general is a very difficult topic which I do not want to discuss at any length. Consider the statement that one motive for my signing a petition was admiration for its promoter, X. Asked ' Why did you sign it?' I might well say ' Well, for one thing, X, who is promoting it, did . . . ' and describe what he did in an admiring way. I might add ' Of course, I know that is not a ground for signing it, but I am sure it was one of the things that most influenced me '—which need *not* mean: ' I thought explicitly of this before signing '. I say ' Consider this ' really with a view to saying ' let us not consider it here '. It is too complicated.

The account of motive popularised by Professor Ryle does not appear satisfactory. He recommends construing ' he boasted from vanity ' as saying ' he boasted . . . and his doing so satisfies the law-like proposition that whenever he finds a chance of securing the admiration and envy of others, he does whatever he thinks will produce this admiration and envy ' [1]. This passage is rather curious and roundabout in expression; it seems to say, and I can't understand it unless it implies, that a man could not be said to have boasted from vanity unless he always behaved vainly, or at least very very often did so. But this does not seem to be true.

To give a motive (of the sort I have labelled ' motive-in-general ', as opposed to backward-looking motives and intentions) is to say something like ' See the action in this light '. To explain one's own actions by an account indicating a motive is to put them in a certain light. This sort of explanation is often elicited by the question ' Why?' The question whether the light in which one so puts one's action is a true light is a notoriously difficult one.

The motives admiration, curiosity, spite, friendship, fear, love of truth, despair and a host of others are either of this extremely complicated kind or are forward-looking or mixed. I call a motive forward-looking if it is an intention. For example, to say that someone did something for fear of . . . often comes to the same as saying he did so lest . . . or in order that . . . should not happen.

14.   Leaving then, the topic of motive-in-general or ' interpretative ' motive, let us return to backward-looking motives. Why is it that in revenge and gratitude, pity and remorse, the past event (or present situation) is a reason for acting, not just a mental cause?

Now the most striking thing about these four is the way in which good and evil are involved in them. E.g. if I am grateful to someone, it is because he has done me some good, or at least I think he has, and I cannot show gratitude by something that I intend to harm him. In remorse, I hate some good things for myself; I could not express remorse by getting myself plenty of enjoyments, or for something that I did not find bad. If I do something out of revenge which is in fact advantageous rather

[1] *The Concept of Mind*, p. 89.

than harmful to my enemy, my action, in its description of being advantageous to him, is involuntary.

These facts are the clue to our present problem. If an action has to be thought of by the agent as doing good or harm of some sort, and the thing in the past as good or bad, in order for the thing in the past to be the reason for the action, then this reason shews not a mental cause but a motive. This will come out in the agent's elaborations on his answer to the question ' Why ? '

It might seem that this is not the most important point, but that the important point is that a *proposed* action can be questioned, and the answer be a mention of something past. ' I am going to kill him '—' Why ? '—' He killed my father '. But if we say this, we show that we are forgetting the course of our enquiry; we do not yet know what a proposed action is; we can so far describe it only as an action predicted by the agent, either without his justifying his prediction at all, or with his mentioning in justification a reason for acting; and the meaning of the expression ' reason for acting ' is precisely what we are at present trying to elucidate. Might one not predict mental causes and their effects ? Or even their effects after the causes have occurred ? E.g. ' This is going to make me angry '. Here it may be worth while to remark that it is a mistake to think one cannot choose whether to act from a motive. Plato saying to a slave ' I should beat you if I were not angry ' would be a case. Or a man might have a policy of never making remarks about a certain person because he could not speak about that man unenviously, or unadmiringly.

We have now distinguished between a backward-looking motive and a mental cause, and found that, here at any rate, what the agent reports in answer to the question ' Why ? ' is a reason for acting if in treating it as a reason he conceives it as something good or bad, and his own action as doing good or harm. If you could e.g. show that either the action for which he has revenged himself, or that in which he has revenged himself, was quite harmless or was beneficial, he ceases to offer a reason, except prefaced by ' I thought '. If it is a proposed revenge he either gives it up or changes his reason. No such discovery would affect an assertion of mental causality. Whether in general good and harm play an essential part in the concept of intention it still remains to find out. So far they have only been introduced as making a clear difference between a backward-looking motive and a mental cause. When the question ' Why? ' about a present action is answered by a description of a future state of affairs, this is already distinguished from a mental cause just by being future. Hence there does not so far seem to be any need to say that intention as such is intention of good or of harm.

15. Now, however, let us consider this case:
　　　 Why did you do it?
　　　 Because he told me to.

Is this a cause or a reason? It appears to depend very much on what the action was or what the circumstances were. And we should often refuse to make any distinction at all between something's being a reason, and its being a cause of the kind in question; for that was explained as what one is after if one asks the agent what led up to and issued in an action. But his being given a reason to act and accepting it might be such a thing. And how would one distinguish between cause and reason in such a case as having hung one's hat on a peg because one's host said ' Hang up your hat on that peg '? Nor, I think, would it be correct to say that this is a reason and not a mental cause because of the understanding of the words that went into accepting the suggestion. Here one would be attempting a contrast between this case and, say, turning round at hearing someone say Boo! But this case would not in fact be decisively on one side or the other; forced to choose between taking the noise as a reason and as a cause, one would probably decide by how sudden one's reaction was. Further, there is no question of understanding a sentence in the following case: ' Why did you waggle your two fore-fingers by your temples? '—' Because *he* was doing it '; but this is not particularly different from hanging one's hat up because one's host said ' Hang your hat up '. Roughly speaking—if one were forced to go on with the distinction—the more the action is described as a mere response, the more inclined one would be to the word ' cause ' ; while the more it is described as a response to something as *having a significance* that is dwelt on by the agent in his account, or as a response surrounded with thoughts and questions, the more inclined one would be to use the word 'reason'. But in very many cases the distinction would have no point.

This, however, does not mean that it never has a point. The cases on which we first grounded the distinction might be called 'full-blown': that is to say, the case of e.g. revenge on the one hand, and of the thing that made one jump and knock a cup off a table on the other. Roughly speaking, it establishes something as a reason if one argues against it; not as when one says 'Noises should not make you jump like that: hadn't you better see a doctor?' but in such a way as to link it up with motives and intentions: 'You did it because he told you to? But why do what he says?' Answers like 'he has done a lot for me', 'he is my father', 'it would have been the worse for me if I hadn't' give the original answer a place among reasons; 'reasons' here of course conforms to our general explanation. Thus the full-blown cases are the right ones to consider in order to see the distinction between reason and cause. But it is worth noticing that what is so commonly said, that reason and cause are everywhere sharply distinct notions, is not true.

16. It will be useful at this stage to summarize conclusions reached so far. Intentional actions are a sub-class of the events in a man's history which are known to him *not* just because he observes them. In this wider class is included one type of involuntary actions, which is marked off by the fact that mental causality is excluded from it; and mental causality is itself characterized by being known without observation. But intentional actions are not marked off just by being subject to mental causality, since there are involuntary actions from which mental causality is not excluded. Intentional actions, then, are the ones to which the question 'Why?' is given application, in a special sense which is so far explained as follows: the question has not that sense if the answer is evidence or states a cause, including a mental cause; positively, the answer may (*a*) simply mention past history, (*b*) give an interpretation of the action, or (*c*) mention something future. In cases (*b*) and (*c*) the answer is already characterised as a reason for acting, i.e. as an answer to the question 'Why?' in the requisite sense; and in case (*a*) it is an answer to that question if the ideas of good or harm are

involved in its meaning as an answer; or again if further enquiry elicits that it is connected with 'interpretative' motive, or intention *with which*.

17. I can now complete my account of when our question 'Why?' is shewn not to apply. We saw that it was refused application if the agent's answer was 'I was not aware I was doing that' and also if the answer implied 'I *observed* that I was doing that'. There was a third circumstance as well, in which the question would have no application: namely that in which the action is somehow characterised as one in which there is no room for what I called mental causality. This would come out if for example the only way in which a question as to cause was dealt with was to speculate about it, or to give reasons why such and such should be regarded as the cause. E.g. if one said 'What made you jump like that?' when someone had just jerked with the spasm which one sometimes gets as one is dropping off to sleep, he would brush aside the question or say 'It was involuntary—you know, the way one does sometimes jump like that'; now a mark of the rejection of that particular question 'What made you?' is that one says things like 'I don't know if anyone knows the cause' or 'Isn't it something to do with electrical discharges?' and that this is the only sense that one gives to 'cause' here.

Now of course a possible answer to the question 'Why?' is one like 'I just thought I would' or 'It was an impulse' or 'For no particular reason' or 'It was an idle action—I was just doodling'. I do not call an answer of this sort a rejection of the question. The question is not refused application because the answer to it says that there is *no* reason, any more than the question how much money I have in my pocket is refused application by the answer 'None'.

An answer of rather peculiar interest is: 'I don't know why I did it'. This can have a sense in which it does not mean that perhaps there is a causal explanation that one does not know. It goes with 'I found myself doing it', 'I heard myself say . . .', but is appropriate to actions in which some special reason seems to be demanded, and one has none. It suggests surprise at one's own actions; but that is not a sufficient condition for saying it, since one can be a bit surprised without wanting to use such an expression—if one has uttered a witticism of a sort that is not one's usual style, for example.

'I don't know why I did it' perhaps is rather often said by people caught in trivial crimes, where however it tends to go with 'it was an impulse'. I disregard this use of it, as it has become too much of a set form; and it does not in fact seem strange to be attracted to commit trivial crimes without any need (if there is anything strange, it is only in not being deterred by obvious considerations, not in thinking of doing such a thing). Sometimes one may say: 'Now why did I do that?'—when one has discovered that, e.g. one has just put something in a rather odd place. But 'I don't know why I did it' may be said by someone who does not *discover* that he did it; he is quite aware as he does it; but he comes out with this expression as if to say 'It is the sort of action in which a reason seems requisite'. As if there were a reason, if only he knew it; but of course that is not the case in the relevant sense; even if psychoanalysis persuades him to accept something as his reason, or he finds a reason in a divine or diabolical plan or inspiration, or a causal explanation in his having been previously hypnotised.

I myself have never wished to use these words in this way, but that does not make me suppose them to be senseless. They are a curious intermediary case: the question 'Why?' has and yet has not application; it has application in the sense that it is admitted as an appropriate question; it lacks it in the sense that the answer is that there is no answer. I shall later be discussing the difference between the intentional and the voluntary; and once that distinction is made we shall be able to say: an action of this sort is voluntary, rather than intentional. And we shall see (§25) that there are other more ordinary cases where the question 'Why?' is not *made out* to be inapplicable, and yet is not granted application.

18. Answers like 'No particular reason'; 'I just thought I would', and so on are often quite intelligible; sometimes strange; and sometimes unintelligible. That is to say, if someone hunted out all the green books in his house and spread them out carefully on the roof, and gave one of these answers to the question 'Why?' his words would be unintelligible unless as

joking and mystification. They would be unintelligible, not because one did not known what *they* meant, but because one could not make out what the man meant by saying them here. These different sorts of unintelligibility are worth dwelling on briefly.

Wittgenstein said that when we call something senseless it is not as it were its sense that is senseless, but a form of words is being excluded from the language. E.g. ' Perhaps congenitally blind people have visual images '. But the argument for ' excluding this form of words from the language ' is apparently an argument that ' its sense is senseless '. The argument goes something like this: What does it mean?—That they have what I have when I have a visual image. And *what* have I?—Something like *this*.—Here Wittgenstein would go on to argue against private ostensive definition. The next move is to see what is the language-game played with ' having a visual image ' or ' seeing in one's mind's eye '. It isn't *just* saying these things—nor can it be explained as saying them with the right reference (this has been shewn by the argument against private ostensive definition). The conclusion is that the language-game with ' seeing ' is a necessary part of the language-game with ' seeing in the mind's eye '; or rather, that a language-game can only be identified as that latter one if the former language-game too is played with the words used. The result of the argument, if it is successful, is that we no longer want to say ' Perhaps blind men . . . etc.' Hence Wittgenstein's talk of ' therapies '. The ' exclusion from the language ' is done not by legislation but by persuasion. The ' sense that is senseless ' is the *type* of sense that our expressions suggest; the suggestion arises from a ' false assimilation of games '.

But our present case is entirely different. If we say ' it does not make sense for this man to say he did this for no particular reason ' we are not ' excluding a form of words from the language'; we are saying ' we cannot understand such a man '. (Wittgenstein seems to have moved from an interest in the first sort of ' not making sense ' to the second as *Philosophical Investigations* developed.)

Similarly, ' I was not aware that I was doing so ' is sometimes intelligible, sometimes strange, and in some cases would be unintelligible.

It would take considerable skill to use language with frequent unintelligibility of this sort; it would be as difficult as to train oneself in the smooth production of long unrehearsed word-salads.

The answers to the question ' Why? ' which give it an application are, then, more extensive in range than the answers which give reasons for acting. This question ' Why? ' can now be defined as the question expecting an answer in this range. And with this we have roughly outlined the area of intentional actions.

19. We do not add anything attaching to the action at the time it is done by describing it as intentional. To call it intentional is to assign it to the class of intentional actions and so to indicate that we should consider the question ' Why? ' relevant to it in the sense that I have described. For the moment, I will not ask *why* this question '. Why? ' should be applicable to some events and not to others.

That an action is not called ' intentional ' in virtue of any extra feature which exists when it is performed, is clear from the following: Let us suppose that there is such a feature, and let us call it ' I '. Now the intentional character of the action cannot be asserted without giving the description under which it is intentional, since the same action can be intentional under one description and unintentional under another. It is however something actually done that is intentional, if there is an intentional action at all. A man no doubt contracts certain muscles in picking up a hammer; but it would generally be false to call his contraction of muscles the intentional act that he performed. This does not mean that his contraction of muscles was unintentional. Let us call it ' preintentional '. Are we to say that I, which is supposed to be the feature in virtue of which what he does is an intentional action, is something which accompanies a preintentional action, or movement of his body? If so, then the preintentional movement $+ I$ guarantees that *an* intentional action is performed: but which one? Clearly our symbol ' I ' must be interpreted as a description, or as having an internal relation to a description, of an action. But nothing about the man considered by himself in the moment of contracting his muscles, and nothing in the contraction of the muscles, can possibly determine the content of that description; which therefore may be *any* one, if we are merely considering what can be determined about the man by himself in the moment. Then it is a mere happy accident that an *I* relevant to the wider context and further consequences *ever* accompanies the preintentional movements in which a man performs a given intentional action. What makes it *true* that the man's movement is one by which he performs such and such an action will have absolutely no bearing on the *I* that occurs, unless we suppose a mechanism by which an *I* appropriate to the situation is able to occur because of the man's knowledge of the situation—he guesses e.g. that his muscular contractions will result in his grasping the hammer and so the right *I* occurs. But that cannot very well be, since a man may very likely not be so much as aware of his preintentional acts. Besides, we surely want *I* to have some effect on what happens. Does he then notice that *I* is followed often enough by its description's coming true, and so summon up *I*? But that turns the summoning up of *I* into an intentional action itself, for which we shall have to look for a second *I*. Thus the assumption that some feature of the moment of acting constitutes actions as intentional leads us into inextricable confusions, and we must give it up.

And in describing intentional actions as such, it will be a mistake to look for *the* fundamental description of what occurs— such as the movements of muscles or molecules—and then think of intention as something, perhaps very complicated, which qualifies this. The only events to consider are intentional actions themselves, and to call an action intentional is to say it is intentional under some description that we give (or could give) of it.

The question does not normally arise whether a man's proceedings are intentional; hence it is often ' odd ' to call them so. E.g. if I saw a man, who was walking along the pavement, turn towards the roadway, look up and down, and then walk across the road when it was safe for him to do so, it would not be usual for me to say that he crossed the road intentionally. But it would be wrong to infer from this that we ought not to give such an action as a typical example of intentional action. It would however be equally a mistake to say: since this man's crossing the road is an example of an intentional action, let us consider this action by itself, and let us try to find in the action, or in the man himself at the moment of acting, the characteristic which makes the action intentional.

20. Would intentional actions still have the characteristic 'intentional', if there were no such thing as expression of intention for the future, or as further intention in acting? I.e. is 'intentional' a characteristic of the actions that have it, which is formally independent of those other occurrences of the concept of intention? To test this, I will make two rather curious suppositions: (*a*) Suppose that 'intention' only occurred as it occurs in 'intentional action', and (*b*) suppose that the only answer to the question 'Why are you X-ing?', granted that the question is not refused application, were 'I just am, that's all'.

(*a*) This supposition, we might say, carries a suggestion that 'intentional action' means as it were 'intentious action'. That is to say, that an action's being intentional is rather like a facial expression's being sad. It would not, of course, be without consequences; the applicability of the question 'Why?' would remain. But of course the diagnosis of a melancholy expression has consequences too, and in a similar fashion: 'What are you sad about?' may be asked, and may receive either a positive answer or the answer 'Nothing'; which in turn may mean that one is sad, but not about anything, or that one is not sad. Intention, on this interpretation of our supposition (*a*), has become a style-characteristic of observable human proceedings, with which is associated the question 'Why?' This however is quite contrary to the concept of intention, because the very same human proceedings may be questioned under the description 'X' ('Why are you X-ing?') and under the description 'Y' ('Why are you Y-ing ?'), and the first question be admitted application while the second is refused it, so that the very same proceedings are intentional under one description and unintentional under another. It is clear that a concept for which this does not hold is not a concept of intention. If we try to make it retain this characteristic by suggesting that the proceedings-in-a-given-description are what bears the stamp of intention, we shall have to suppose that a man who, having been seen clearly, is asked 'Why are you X-ing?' can never profess unawareness that he was X-ing, except on pain of being a liar if in fact he was X-ing.

And this supposition would involve such radical changes that it becomes impossible to say whether we could still see a place for the concept of intention at all, or diagnose the question ' Why? ' as having in part the same sense as our question ' Why? ' We should merely have *a* question to which possible answers were ' I just was, that's all ', ' I wasn't ', mention of something in the past like ' He killed my father ', or a sentimental characterisation of the action. For of course answers giving further intentions are excluded *ex hypothesi*, since if they were included the possible substitutions for ' X ' in 'A intends X ' would include more than the supposition allows.

We can however try to give a different interpretation to supposition (*a*). Intention still only occurs in present action. That is, there is still no such thing as the further intention *with* which a man does what he does; and no such thing as intention for the future. Intention however is not a style that marks an action, or an action-in-a-description; for it is possible for a man to think he is doing one thing when he is not doing that thing but another. Thus he can say that he did not know he was doing something, when asked why he did it. We must not however be too sweeping in excluding intention *with* which a man does what he does; for we must presumably allow the further intention with which he is doing X, say Y, so long as it is reasonable to say that he is doing Y in, and at the same time as, doing X: e.g. a man can be said to hold a glass to his lips with (at least) the intention of drinking, if he *is* drinking when he holds it to his lips. What is excluded from the supposition is a further intention Y such that we *could* object that he is not yet doing Y but only doing X with a view to doing Y, as when a man takes his gun down with a view to shooting rabbits.

In this case intentional actions will be marked out as those of which a man has non-observational knowledge, and for which there is a question whose answers fall in the range (*a*) ' I just did ' (*b*) backward looking motive, and (*c*) sentimental characterisation. (*a*) is of no interest; so our question must be: is *motive* enough to constitute intentional actions as a special kind? One can argue against motives—i.e. criticise a man for having acted on such a motive—but a great deal of the point of doing so will be gone if we imagine the expression of intention for the future to be absent, as it is on our hypothesis. That is why on this hypothesis giving an interpretative motive turns into sentimental character-isation. It seems reasonable to say that if the only occurence of intention were as the intention of doing whatever one is doing, the notion of intentional action itself would be a very thin one; it is not clear why it should be marked off as a special class among all those of a man's actions and movements which are known to him without observation, any more than we mark off movements that are expressions of emotion as a distinct and important class of happenings.

(*b*) By the second supposition, though intention is supposed to occur both in present intentional action and in expressions of intention for the future, the only answer to the question 'Why?' is 'I just am'. (Naturally 'further intention with which' a man acts is excluded by this hypothesis, for it is expressed in a type of answer to the question 'Why?' which is excluded.) If this were so, then there would be no special sense of the question 'Why?' and no distinct concept of intentional action at all. That is to say, it would no longer be possible to differ-entiate within the class of acts known without observation. For a question whose only answer is a statement that one *is* doing the thing cannot be identified with our question 'Why?', even if the word for it is one used in requests for evidence and enquiries into causality. Thus on the present hypothesis there would be no distinction between such things as starts and gasps and, quite generally, *voluntary* actions.

It is natural to think that the difference is one that we can see in the things themselves. To be sure, all these things will be alike as regards the way we know that they are taking place—but isn't there an introspectively discernible difference between an involuntary gasp and a voluntary intake of breath?—Well, one may be more sudden than the other. Still, I can voluntarily do it quite quickly, so that is not the difference.—Should we say the voluntary kind can be *foreseen*, predicted?—But the involun-tary kind might be predicted.—But the *basis* of the prediction won't be the same!—To be sure; but the difference between bases of prediction is just the difference between evidence and a reason for acting. Though 'I just did, that's all' is an answer to the question 'Why did you do it?', it does not give a reason, and the parallel answer for the future ' I'm just going to, that's all ' does not give a basis for the prediction, it merely repeats it.

Let us try another method of differentiation. A voluntary action can be commanded. If someone says ' Tremble ' and I tremble I am not *obeying* him—even if I tremble because he said it in a terrible voice. To play it as obedience would be a kind of sophisticated joke (characteristic of the Marx Brothers) which might be called ' playing language-games wrong '. Now we can suppose that human actions, which are not distinguished by the way their agent knows them, are or are not subject to command. If they are subject to command they can be distinguished as a separate class; but the distinction seems to be an idle one, just made for its own sake. Don't say ' But the distinction relates to an obviously *useful* feature of certain actions, namely that one can get a person to perform them by commanding him '; for ' usefulness ' is not a concept we can suppose retained if we have done away with ' purpose '.

Still, some actions are subject to command, so has not the question ' Why? ' a place? ' Why did you do it? ' ' Because you told me to '. That is an answer, and if some actions were subject to command, the people concerned might have the question whether something was done in obedience to a command or not. But the question ' Why? ' may here simply be rendered by ' Commanded or not commanded?' This will be a form of the relevant question ' Why? ' if it is open to the speaker to say ' You commanded it, and I did it, but not commanded '. ' I didn't do it because you told me to '.) But what would be the point of this, taken by itself—i.e. in isolation from a person's reasons and aims? For these are excluded; the question ' Why? ' is not supposed to have any such application in the case we are imagining. The expression might be only a form of rudeness.

Thus the occurrence of other answers to the question ' Why? ' besides ones like ' I just did ', is essential to the existence of the concept of an intention or voluntary action.

21. Ancient and medieval philosophers—or some of them at any rate—regarded it as evident, demonstrable, that human beings must always act with some end in view, and even with some one end in view. The argument for this strikes us as rather

D

strange. Can't a man just do what he does, a great deal of the time? He may or may not have a reason or a purpose; and if he has a reason or purpose, it in turn may just be what he happens to want; why demand a reason or purpose for *it*? and why must we at last arrive at some *one* purpose that has an intrinsic finality about it? The old arguments were designed to show that the chain could not go on for ever; they pass us by, because we are not inclined to think it *must* even begin; and it can surely stop where it stops, no need for it to stop at a purpose that looks intrinsically final, one and the same for all actions. In fact there appears to be an illicit transition in Aristotle, from ' all chains must stop somewhere ' to ' there is somewhere where all chains must stop.'

But now we can see why *some* chain must at any rate begin. As we have seen, this does not mean that an action cannot be called voluntary or intentional unless the agent has an end in view; it means that the concept of voluntary or intentional action would not exist, if the question ' Why? ', with answers that give reasons for acting, did not. Given that it does exist, the cases where the answer is ' For no particular reason ', etc. can occur; but their interest is slight, and it must not be supposed that because they can occur that answer would everywhere be intelligible, or that it could be the only answer ever given.

22. In all this discussion, when I have spoken of the answer to the question ' Why? ' as mentioning an *intention*, the intention in question has been of course the intention *with which* a man does what he does. We must now turn to the closer examination of this. So far I have merely said ' If the answer to the question ' Why? ' is a simple mention of something future, then it expresses the intention ', and the question of cause *versus* reason, which has plagued us in relation to answers mentioning the past, simply does not arise here. I do not of course mean to say that every answer which tells you with what intention a man is doing whatever it is he is doing is a description of some future state of affairs; but if a description of some future state of affairs makes sense just by itself as an answer to the question, then it is an expression of intention. But there are other expressions of the intention with which a man is doing something: for example, a

wider description of *what* he is doing. For example, someone comes into a room, sees me lying on a bed and asks ' What are you doing? ' The answer ' lying on a bed ' would be received with just irritation; an answer like ' Resting ' or ' Doing Yoga ', which would be a description of what I am doing in lying on my bed, would be an expression of intention.

For the moment, however, let me concentrate on the simple future answer. I have said an answer describing something future ' just by itself ' is an expression of the intention with which a person acts. That qualification is necessary can be seen in the following instance ' Why are you setting up a camera on this pavement?' ' Because Marilyn Monroe is going to pass by '. That is just a statement of something future, but by no means expresses that I am setting up a camera with the intention that Marilyn Monroe shall pass by. On the other hand, if you say ' Why are you crossing the road ' and I reply ' I am going to look in that shop window ', this expresses the intention with which I cross the road. Now what is the difference?

Consider this case: ' Why are you crossing the road?'— ' Because there will be an eclipse in July '. This answer, as things are, needs filling in. And no kind of filling in that *we* shall accept without objection would give that answer the role of a statement of intention. (I mean e.g. something like ' For six months before the eclipse that shop window is having a lot of explanatory diagrams and models on display '). But some savage might well do something in order to procure an eclipse; and I suppose the answer 'Eclipse in July' could perhaps have been understood as an expression of intention by the Dublin crowd who once assembled to watch an eclipse, and dispersed when Dean Swift sent down his butler with a message to say that by the Dean's orders the eclipse was off.

That is to say: the future state of affairs mentioned must be such that we can understand the agent's thinking it will or may be brought about by the action about which he is being questioned.

But does this mean that people must have notions of cause and effect in order to have intentions in acting? Consider the question ' Why are you going upstairs? ' answered by ' To get my camera '. My going upstairs is not a cause from which anyone could deduce the effect that I get my camera. And yet isn't it a future state of affairs which *is going to be* brought about by my going upstairs? But who can say that it is going to be brought about? Only I myself, in this case. It is not that going upstairs usually produces the fetching of cameras, even if there is a camera upstairs—unless indeed the context includes an order given me, ' Fetch your camera ', or my own statement ' I am going to get my camera '.

On the other hand, if someone says ' But your camera is in the cellar ', and I say ' I know, but I am still going upstairs to get it ' my saying so becomes mysterious; at least, there is a gap to fill up. Perhaps we think of a lift which I can work from the top of the house to bring the camera up from the bottom. But if I say: ' No, I quite agree, there is no way for a person at the top of the house to get the camera; but still I am going upstairs to get it ' I begin to be unintelligible. In order to make sense of ' I do P with a view to Q ', we must see how the future state of affairs Q is supposed to be a possible later stage in proceedings of which the action P is an earlier stage. It is true that, on the one hand, cases of scientific knowledge, and on the other hand cases of magical rites, or of a vague idea of great power and authority like Dean Swift's, all come under this very vague and general formula. *All* that I have said, in effect, is ' It is not the case that a description of *any* future state of affairs can be an answer to this question about a present action '. A man's intention in acting is not so private and interior a thing that he has absolute authority in saying *what* it is—as he has absolute authority in saying *what* he dreamt. (If what a man says he dreamed does not make sense, that doesn't mean that his saying he dreamed it does not make sense.)

I shall not try to elaborate my vague and general formula, that we must have an idea how a state of affairs Q is a stage in proceedings in which the action P is an earlier stage, if we are to be able to say that we do P so that Q. For of course it is not necessary to exercise these general notions in order to say ' I do P so that Q '. All that it is necessary to understand is that to say, in one form or another: ' But Q won't happen, even if you do P ', or 'but it will happen whether you do P or not' is, in some way, to contradict the intention.

23. Let us ask: is there any description which is *the* description of an intentional action, given that an intentional action occurs? And let us consider a concrete situation. A man is pumping water into the cistern which supplies the drinking water of a house. Someone has found a way of systematically contaminating the source with a deadly cumulative poison whose effects are unnoticeable until they can no longer be cured. The house is regularly inhabited by a small group of party chiefs, with their immediate families, who are in control of a great state; they are engaged in exterminating the Jews and perhaps plan a world war.—The man who contaminated the source has calculated that if these people are destroyed some good men will get into power who will govern well, or even institute the Kingdom of Heaven on earth and secure a good life for all the people; and he has revealed the calculation, together with the fact about the poison, to the man who is pumping. The death of the inhabitants of the house will, of course, have all sorts of other effects; e.g., that a number of people unknown to these men will receive legacies, about which they know nothing.

This man's arm is going up and down, up and down. Certain muscles, with Latin names which doctors know, are contracting and relaxing. Certain substances are getting generated in some nerve fibres—substances whose generation in the course of voluntary movement interests physiologists. The moving arm is casting a shadow on a rockery where at one place and from one position it produces a curious effect as if a face were looking out of the rockery. Further, the pump makes a series of clicking noises, which are in fact beating out a noticeable rhythm.

Now we ask: What is this man doing? What is *the* description of his action?

First, of course, *any* description of what is going on, with him as subject, which is in fact true. E.g. he is earning wages, he is supporting a family, he is wearing away his shoe-soles, he is making a disturbance of the air. He is sweating, he is generating those substances in his nerve fibres. If in fact good government, or the Kingdom of Heaven on earth and a good life for everyone, comes about by the labours of the good men who get into power because the party chiefs die, then he will have been helping to produce this state of affairs. However, our enquiries into the question ' Why? ' enable us to narrow down our consideration of descriptions of what he is doing to a range covering all and only his intentional actions. ' He is X-ing ' is a description of an intentional action if (*a*) it is true and (*b*) there is such a thing as an answer in the range I have defined to the question ' Why are you X-ing? ' That is to say, the description in ' Why are you contracting those muscles? ' is ruled out if the *only* sort of answer to the question ' Why? ' displays that the man's knowledge, if any, that he was contracting those muscles is an inference from his knowledge of anatomy. And the description in the question ' Why are you generating those substances in your nerve fibres? ' will *in fact* always be ruled out on these lines unless we suppose that the man has a plan of producing these substances (if it were possible, we might suppose he wanted to collect some) and so moves his arm vigorously to generate them. But the descriptions in the questions ' Why are you making that face come and go in the rockery? ', ' Why are you beating out that curious rhythm? ' will be revealed as descriptions of intentional actions or not by different styles of answer, of which one would contain something signifying that the man *notices* that he does that, while the other would be in the range we have defined. But there are a large number of X's, in the imagined case, for which we can readily suppose that the answer to the question ' Why are you X-ing? ' falls within the range. E.g. ' Why are you moving your arm up and down? '—' I'm pumping '. ' Why are you pumping? '—' I'm pumping the water-supply for the house '. ' Why are you beating out that curious rhythm? ',—' Oh, I found out how to do it, as the pump does click anyway, and I do it just for fun '. 'Why are you pumping the water? '—' Because it's needed up at the house ' and (*sotto voce*) ' To polish that lot off '. ' Why are you poisoning these people? '—' If we can get rid of them, the other lot will get in and . . . '

Now there is a break in the series of answers that one may get to such a question. Let the answer contain a further description Y, then sometimes it is correct to say not merely: the man is X-ing, but also: ' the man is Y-ing '—if that is, nothing falsifying the statement ' He is Y-ing ' can be observed. E.g. ' Why are you pumping ? '—' To replenish the water supply '. If this was the answer, then we can say ' He *is* replenishing the water-supply '; unless indeed, he is not. This will appear a tautologous pronouncement; but there *is* more to it. For if after his saying ' To replenish the water-supply ' we can say ' He is replenishing the water-supply ', then this would, in ordinary circumstances, of itself be enough to characterise *that* as an intentional action. (The qualification is necessary because an intended effect just occasionally comes about by accident). Now that is to say, as we have already determined, that the same question ' Why? ' will have application to this action in its turn. This is not an empty conclusion: it means that someone who, having so answered ' To replenish the water-supply ', is asked ' Why are you replenishing the water-supply? ', must not say e.g. ' Oh, I didn't know I was doing that ', or refuse any but a causal sense of the question. Or rather, that if he does, this makes nonsense of his answers.

A man can *be doing* something which he nevertheless does not *do*, if it is some process or enterprise which it takes time to complete and of which therefore, if it is cut short at any time, we may say that he *was doing* it, but *did not do* it. This point however, is in no way peculiar to intentional action; for we can say that something was falling over but did not fall (since something stopped it). Therefore we do not appeal to the presence of intention to justify the description ' He is Y-ing '; though in some cases his own statement that he is Y-ing may, at a certain stage of the proceedings, be needed for anybody else to be able to say he is Y-ing, since not enough has gone on for that to be evident; as when we see a man doing things with an array of wires and plugs and so on.

Sometimes, jokingly, we are pleased to say of a man ' He is doing such-and-such ' when he manifestly is not. E.g. ' He is replenishing the water-supply ', when this is not happening because, as we can see but he cannot, the water is pouring out of a hole in a pipe on the way to the cistern. And in the same way we may speak of some rather doubtful or remote objective, e.g. ' He is proving Fermat's last theorem '; or again one might say of a madman ' He is leading his victorious armies '. It is easy, however, to exclude these cases from consideration and point out the break between cases where we can say ' He is Y-ing ', when he has mentioned Y in answer to the question ' Why are you X-ing? ', and ones where we say rather ' He is going to Y '. I do not think it is a quite sharp break. E.g. is there much to choose between ' She is making tea ' and ' She is putting on the kettle in order to make tea '—i.e. ' She is going to make tea' ? Obviously not. And hence the common use of the present to describe a future action which is by no means just a later stage in activity which has a name as a single whole. E.g. ' I am seeing my dentist ', ' He is demonstrating in Trafalgar Square ' (either might be said when someone is at the moment e.g. travelling in a train). But the less normal it would be to take the achievement of the objective as a matter of course, the more the objective gets expressed *only* by ' in order to '. E.g. ' I am going to London in order to make my uncle change his will ' ; not ' I am making my uncle change his will '.

To a certain extent the three divisions of the subject made in §1, are simply equivalent. That is to say, where the answers ' I am going to fetch my camera ', ' I am fetching my camera ' and ' in order to fetch my camera ' are interchangeable as answers to the question ' Why ? ' asked when I go upstairs.

Now if all this holds, what are we to say about all these many descriptions of an intentional action? Are we to say that there are as many distinct actions as we can generate distinct descriptions, with X as our starting point? I mean: We say ' Why are you X-ing? ' and get the answer ' To Y ', or ' I'm Y-ing ', Y being such that we can say ' he's Y-ing '; and then we can ask ' Why are you Y-ing? ' and perhaps get the answer ' To Z ', and can still say ' He's Z-ing '. E.g. ' Why are you moving your arm up and down? ' ' To operate the pump ', and he is operating the pump. ' Why are you pumping? ' ' To replenish the water-supply ' and he is replenishing the water-supply; ' Why are you replenishing the water-supply? ' ' To poison the inhabitants ' and he is poisoning the inhabitants, for they are getting poisoned. And here comes the break; for though in the case we have described there is probably a further answer, other than ' just for fun ', all the same this further description (e.g. to save the Jews, to put in the good men, to get the Kingdom of Heaven on earth) is not such that we can now say: he is saving the Jews, he is getting the Kingdom of Heaven, he is putting in the good ones. So let us stop here and say: are there four actions here, because we have found four distinct descriptions satisfying our conditions, namely moving his arm up and down, operating the pump, replenishing the water supply, and poisoning the inhabitants?

24. Before trying to answer this, however, we must raise some difficulties. For someone might raise the objection that pumping can hardly be an act of poisoning. It is of course, as the lawyers would say, an act of laying poison, and one might try to reply by saying the man poisons the inhabitants if he lays poison and they get poisoned. But after all we said it was a cumulative poison; this means that no single act of laying the poison is by itself an act of poisoning; besides, didn't the other man ' lay ' the poison? Suppose we ask ' When did our man poison them? ' One might answer: all the time they got poisoned. But in that case one might say ' His poisoning them was not an action; for he was perhaps doing nothing relevant at any of the times they were drinking the poison.' Is the question ' When exactly did he poison them?', to be answered by specifying all the numerous times when he laid the poison? But none of them by itself could be called poisoning them; so how can we call the man's present pumping an intentional act of poisoning? Or must we draw the conclusion that he at no time poisoned them, since he was not engaged in poisoning at the times at which they were being poisoned? We cannot say that since at some time he poisoned them, there *must* be actions which we can label ' poisoning them ', and in which we can find what it was to poison them. For in the acts of pumping poisoned water nothing in particular is necessarily going on that might not equally well have been going on if the acts had been acts of pumping non-poisonous water. Even if you imagine that pictures of the inhabitants lying dead occur in the man's head, and please him— such pictures could also occur in the head of a man who was *not* poisoning them, and *need* not occur in this man. The difference appears to be one of circumstances, not of anything that is going on *then*.

25. A further difficulty however arises from the fact that the man's intention might not be to poison them but only to earn his pay. That is to say, if he is being improbably confidential

and is asked 'Why did you replenish the house water-supply with poisoned water?', his reply is, not 'To polish them off', but 'I didn't care about that, I wanted my pay and just did my usual job'. In that case, although he knows concerning an intentional act of his—for it, namely replenishing the house water-supply, is intentional by our criteria—that it is *also* an act of replenishing the house water-supply with *poisoned* water, it would be incorrect, by our criteria, to say that his act of replenishing the house supply with poisoned water was intentional. And I do not doubt the correctness of the conclusion; it seems to shew that our criteria are rather good. On the other hand, we really do seem to be in a bit of a difficulty to find the intentional act of poisoning those people, supposing that this is what his intentional act is. It is really not at all to be wondered at that so very many people have thought of intention as a special interior movement; then the thing that marked this man's proceedings as *intentional* poisoning of those people would just be that this interior movement occurred in him. But (quite apart from the objections to this idea which we have already considered) the notion of the interior movement tends to have the most unfortunately absurd consequences. For after all we can *form* intentions; now if intention is an interior movement, it would appear that we can choose to have a certain intention and not another, just by e.g. saying within ourselves: 'What I *mean* to be doing is earning my living, and *not* poisoning the household'; or 'What I *mean* to be doing is helping those good men into power; I withdraw my intention from the act of poisoning the household, which I prefer to think goes on without my intention being in it'. The idea that one can determine one's intentions by making such a little speech to oneself is obvious bosh. Nevertheless the genuine case of 'I didn't care tuppence one way or the other for the fact that someone had poisoned the water, I just wanted to earn my pay without trouble by doing my usual job—I go with the house, see? and it doesn't matter to me who's in it' does appear to make it very difficult to find anything except a man's thoughts—and these are surely interior—to distinguish the intentional poisoning from poisoning knowingly when this was nevertheless not the man's intention.

Well, one may say, isn't my proposed criterion in a way a

criterion by thoughts? If the answer to the question ' Why did you replenish the house supply with poisoned water? ' is ' To polish them off ', or any answer within the range, like ' I just thought I would ', then by my criterion the action under that description is characterised as intentional; otherwise not. But does this not suppose that the answer is or would be *given*? And a man can surely make up the answer that he prefers! So it may appear that I have supplied something just like the interior movement, which a man can make what he likes; but (perhaps out of an attachment to ' verificationism ') preferred an external answer (actual or hypothetical) which a man can equally make what he likes—at least within the range of moderately plausible answers. Of course I must mean that the *truthful* answer is, or would be, one or the other; but what sort of control of truthfulness can be established here?

The answer to this has to be: there can be a certain amount of control of the truthfulness of the answer. For example, in the case of the man who didn't care tuppence, part of the account we imagined him as giving was that he just went on doing his usual job. It is therefore necessary that it should be his usual job if his answer is to be acceptable; and he must not do anything, out of the usual course of his job, that assists the poisoning and of which he cannot give an acceptable account. E.g. suppose he distracts the attention of one of the inhabitants from something about the water source that might suggest the truth; the question ' Why did you call him from over there? ' must have a credible answer other than ' to prevent him from seeing'; and a multiplication of such points needing explanation would cast doubt on his claim not to have done anything with a view to facilitating the poisoning.—And yet here we might encounter the following explanation: he did not want the enormous trouble that would result from a certain person's noticing; hoped that since the poison was laid it would all go off safely. All along the line he calculated what looked like landing him personally in least trouble, and he reckoned that preventing anything from being suspected would do that. That is quite possible.

Up to a point, then, there is a check on his truthfulness in the account we are thinking he would perhaps give; but still, there is an area in which there is none. The difference between the cases in which he doesn't care whether the people are actually poisoned or not, and in which he is very glad on realising that they will be poisoned if he co-operates by going on doing his ordinary job, is not one that necessarily carries with it any difference in what he overtly does or how he looks. The difference in his thought on the subject *might* only be the difference between the meanings of the grunt that he gives when he grasps that the water is poisoned. That is to say, when asked ' Why did you replenish the house supply with poisoned water? ' he might either reply ' I couldn't care tuppence ' or say ' I was glad to help to polish them off ', and if capable of saying what had actually occurred in him at the time as the vehicle of either of these thoughts, he might have to say only that he grunted. This is the kind of truth there is in the statement ' Only you can know if you had such-and-such an intention or not '. There is a point at which only what the man himself says is a sign; and here there is room for much dispute and fine diagnosis of his genuineness.

On the other hand, if, say, this was not his normal job, but he was hired by the poisoner to pump the water, knowing it was poisoned, the case is different. He can say he doesn't care tuppence, and that he only wants the money; but the commission by the acceptance and performance of which he gets the money is—however implicit this is allowed to be—to pump poisoned water. Therefore unless he takes steps to cheat his hirer (he might e.g. put what he mistakenly thought was an antidote into the water), it is not an acceptable account if he says ' I wasn't intending to pump poisoned water, only to pump water and get my hire ', so that the forms he adopts for refusing to answer the question ' Why did you pump poisoned water? ' with an answer in our defined range—e.g. with the answer ' to get the pay '— are unacceptable. So that while we can find cases where ' only the man himself can say whether he had a certain intention or not; they are further limited by this: he cannot profess not to have had the intention of doing the thing that was a means to an end of his.

All this, I think, serves to explain what Wittgenstein says at §644 of *Philosophical Investigations*:

    ' " I am not ashamed of what I did then, but of the intention which I had ". And didn't the intention reside *also* in what I did? What justifies the shame? The whole history of the incident.'

And against the background of the qualifications we have introduced, we can epitomize the point by saying 'Roughly speaking, a man intends to do what he does'. But of course that is *very* roughly speaking. It is right to formulate it, however, as an antidote against the absurd thesis which is sometimes maintained: that a man's intended action is only described by describing his *objective*.

The question arises: what can be the interest of the intention of the man we have described, who was only doing his usual job, etc.? It is certainly not an ethical or legal interest; if what he said was true, *that* will not absolve him from guilt of murder! We just *are* interested in what is true about a man in this kind of way. Here again Wittgenstein says something relevant, in his discussion of ' I was going to ':

> ' Why do I want to tell him about an intention too, as well as telling him what I did? . . . because I want to tell him something about *myself*, which goes beyond what happened at that time. I reveal to him something of myself when I tell him what I was going to do.—Not, however, on grounds of self-observation, but by way of a response (it might also be called an intuition).'
>
> (*Philosophical Investigations*, §659).

Wittgenstein is presumably thinking of a response, or reaction, to the memory of ' that time '; in the context of *our* interests, we can think of it as a response to our special question ' Why? '.

26. Let us now return to the question with which we ended §23: Are we to say that the man who (intentionally) moves his arm, operates the pump, replenishes the water supply, poisons the inhabitants, is performing *four* actions? Or only one? The answer that we imagined to the question ' Why? ' brings it out that the four descriptions form a series, A—B—C—D, in which each description is introduced as dependent on the previous one, though independent of the following one. Then is B *a* description of A, C of B, and so on? Not if that means that we can see that ' he is operating the pump ' is another

description of what is here also described by ' he is moving his arm up and down '—in such a way that is, that what verifies the latter, in this case, also verifies the former. On the other hand, if we say there are four actions, we shall find that the only *action* that B consists in here is A; and so on. Only, more circumstances are required for A to be B than for A just to be A. And far more circumstances for A to be D, than for A to be B. But these circumstances *need* not include any particularly recent action of the man who is said to do A, B, C and D (although we made it a cumulative poison, for present purposes we can suppose that a single pumping is enough to do the trick). In short, the only distinct action of his that is in question is this one, A. For moving his arm up and down with his fingers round the pump handle *is*, in these circumstances, operating the pump; and, in these circumstances, it *is* replenishing the house water-supply; and, in these circumstances, it *is* poisoning the household.

So there is one action with four descriptions, each dependent on wider circumstances, and each related to the next as description of means to end; which means that we can speak equally well of *four* corresponding intentions, or of *one* intention—the last term that we have brought in in the series. By making it the last term so far brought in, we have given it the character of being the intention (so far discovered) *with* which the act in its other descriptions was done. Thus when we speak of four intentions, we are speaking of the character of being intentional that belongs to the act in each of the four descriptions; but when we speak of one intention, we are speaking of intention *with which*; the last term we give in such a series gives the intention *with* which the act in each of its other descriptions was done, and this intention so to speak swallows up all the preceding intentions *with* which earlier members of the series were done. The mark of this ' swallowing up ' is that it is not wrong to give D as the answer to the question ' Why? ' about A; A's being done with B as intention does not mean that D is only indirectly the intention of A, as, if I press on something which is pressing on something . . . which is pressing against a wall, I am only indirectly pressing against the wall. If D is given as the answer to the question ' Why? ' about A, B and C can make an appearance in answer to a question ' How? '. When terms are related in this fashion, they constitute a series of means, the last term of which is, just by being given as the last, so far treated as end.

A term falling outside the series A—D may be a term in another series with some of the members A, B, C in it: for example, if the man is beating out the rhythm of God Save the King in the clicking of the pump. The intention of doing so *with* which he moves his arm up and down is not ' swallowed up ' by the intention of D (beating out that rhythm is not *how* he pumps the water); and the mark of this is that if the question ' Why are you moving your arm up and down? ' receives as answer ' To click out the rhythm of God Save the King ', the answer to ' Why? ' asked about *this* action does not lead to D.

Another implication of what I call ' swallowing up ' is that nothing definite has to hold about *how many* terms we put between A and D; for example, in the imagined case we did not put in a term ' making the water flow along the pipes ', which yet would take its place in the series if anyone thought of asking the question ' Why? ' about it.

27. Is there ever a place for an interior act of intention? I suppose that the man I imagined, who said ' I was only doing my usual job ', might find this formula and administer it to himself in the present tense at some stage of his activities. However, if he does this, we notice that the question immediately arises: with what intention does he do it? This question would always arise about anything which was deliberately performed as an ' act of intending '. The answer in this case might be ' So that I don't have to consider whose side I am on '. Thus the interior performance has not secured what you might have thought, namely that the man's action in pumping the water *is* just doing his usual job; it is itself a new action, like clicking out the rhythm of God Save the King on the pump. It is in fact only if the thought ' I'm only doing my usual job ' is spontaneous rather than deliberate that its occurrence has some face-value relevance to the question what the man's intentions really are. And when spontaneous, it is subject to those tests for truthfulness, which, as we saw, applied to the same form of words given as an explanation after the event; and given that it survives all the same external tests, it comes under the same last deter-

mination: ' *In the end* only you can know whether that is your intention or not '; that means only: there comes a point where a man can say ' This is my intention ', and no one else can contribute anything to settle the matter. (It does not mean that when he says ' This is my intention ', he is evincing a knowledge available only to him. I.e. here ' knows ' only means ' can say '. Unless indeed we imagine a case where it could be said: he *thought* this was his intention, but it became clear that he was deceived.) The only new possibility would be one of eliciting some obviously genuine reaction by saying such things as (to give crude examples): ' Well, then you won't be much interested to hear that the poison is old and won't work '; or ' Then you won't be claiming a share in a great sum with which someone wishes to reward the conspirators '. This sort of thing is of course a stock way of bringing out pretences, often met with in literature—e.g. the deaf man who hears clearly what he ought not to—and in life pretences are no doubt discerned by skilled psychological detectives. But there comes a point at which the skill of psychological detectives has no criteria for its own success. For, after all, probing questions may lead a man to pretend something new, instead of revealing what was there already. So perhaps no concrete inferences as to matters of fact which are quite simply testable can be drawn from the detectives' verdicts. One may *feel* that the verdict is right; that the man who gives it has ' insight '. But, as Wittgenstein put it (*Philosophical Investigations*, p. 128) the consequences here are of a diffuse kind. ' The difference of attitude that one has ' would be a diffuse consequence; or if you want ' consequence ' to mean ' inference ', the nuances in relationships with others in the plot that you will expect the man to have later; the atmosphere between him and them, and similar things.

We can imagine an intention which is a purely interior matter nevertheless changing the whole character of certain things. A contemptuous thought might enter a man's mind so that he meant his polite and affectionate behaviour to someone on a particular occasion only ironically, without there being any outward sign of this (for perhaps he did not venture to give any outward sign). There need not be any specific history, or any consequences, in the light of which an outside observer could see the forms of affection as ironically meant; for as far as concerns history, it is always possible to find things to despise in people without any very special story issuing in contempt on this occasion; and afterwards he might change his mind, think of the episode as an odd aberration, and never turn future occasions into a development of it. Let us suppose that the thought in his mind is ' you silly little twit! ' Now here too, it is not enough that these words should occur to him. He has to mean them. This shews once more, that you cannot take any performance (even an interior performance) as itself an act of intention; for if you describe a performance, the fact that it has taken place is not a proof of intention; words for example may occur in somebody's mind without his meaning them. So intention is never a performance in the mind, though in some matters a performance in the mind which is seriously *meant* may make a difference to the correct account of the man's action—e.g. in embracing someone. But the matters in question are necessarily ones in which outward acts are ' significant ' in some way.

28. We must now look more closely into the formula which has so constantly occurred in this investigation: ' known without observation '. This had its first application to the position of one's limbs and certain movements, such as the muscular spasm in falling asleep. It is not ordinarily possible to find anything that shows one that one's leg is bent. It may indeed be that it is because one has sensations that one knows this; but that does not mean that one knows it by identifying the sensations one has. With the exterior senses it is usually possible to do this. I mean that if a man says he saw a man standing in a certain place, or heard someone moving about, or felt an insect crawling over him, it is possible at least to ask whether he misjudged an appearance, a sound, or a feeling; that is, we can say: Look, isn't *this* perhaps what you saw? and reproduce a visual effect of which he may say: ' Yes, that is, or could be, what I saw, and I admit I can't be sure of more than that '; and the same with the sound or the feeling.[1]

---

[1] I think that these facts ought to make people less contemptuous of phenomenalism than it has now been fashionable to be for a good many years; I have heard people jeer at the expression ' seeing an appearance ' on the grounds that it is incorrect speech. It does not seem to me to matter whether it is incorrect speech or not; the fact remains that one can distinguish between actually seeing a man, and

E

But with e.g. the position of one's limbs it is otherwise than with the external senses. If a man says that his leg is bent when it is lying straight out, it would be incorrect to say that he had misjudged an inner kinaesthetic appearance as an appearance of his leg bent, when in fact what was appearing to him was his leg stretched out. (This topic is certainly a difficult one, deserving a fuller discussion; here, however, such a discussion would be out of place). This consideration, assuming its correctness, is enough to justify saying that normally one does not know the position or movement of one's limbs ' by observation '.

In enquiring into intentional action, however, I have used this formula quite generally, and the following objection will very likely have occurred to a reader: ' Known without observation ' may very well be a justifiable formula for knowledge of the position and movements of one's limbs, but you have spoken of all intentional action as falling under this concept. Now it may be e.g. that one paints a wall yellow, meaning to do so. But is it reasonable to say that one ' knows without observation ' that one is painting a wall yellow? And similarly for all sorts of actions: any actions that is, that are described under any aspect beyond that of bodily movements.

My reply is that the topic of an intention may be matter on which there is knowledge or opinion based on observation, inference, hearsay, superstition or anything that knowledge or opinion ever are based on; or again matter on which an opinion is held without any foundation at all. When knowledge or opinion are present concerning what is the case, and what can happen—say Z—if one does certain things, say ABC, then it is possible to have the intention of doing Z in doing ABC; and if the case is one of knowledge or if the opinion is correct, then doing or causing Z is an intentional action, and it is not by observation that one knows one is doing Z; or in so far as one is observing, inferring etc. that Z is actually taking place, one's knowledge is not the knowledge that a man has of his intentional actions. By the knowledge that a man has of his intentional

the appearances' being such that one says one is seeing, or saw, a man; and that one can describe or identify ' what one saw ' on such an occasion without knowing e.g. that one really saw a reflection of oneself or a coat hanging on a hook; now when one does so describe or identify ' what one saw ', it is perfectly reasonable to call this: describing or identifying an appearance.

actions I mean the knowledge that one denies having if when asked e.g. ' Why are you ringing that bell? ' one replies ' Good heavens! I didn't know *I* was ringing it! '

This is difficult. Say I go over to the window and open it. Someone who hears me moving calls out: What are you doing making that noise? I reply ' Opening the window '. I have called such a statement knowledge all along; and precisely because in such a case what I say is true—I do open the window; and that means that the window is getting opened by the movements of the body out of whose mouth those words come. But I don't say the words like this: ' Let me see, what is this body bringing about? Ah yes! the opening of the window '. Or even like this ' Let me see, what are my movements bringing about? The opening of the window '. To see this, if it is not already plain, contrast this case with the following one: I open the window and it focuses a spot of light on the wall. Someone who cannot see me but can see the wall, says ' What are you doing making that light come on the wall? ' and I say 'Ah yes, it's opening the window that does it ', or ' That always happens when one opens that window at midday if the sun is shining.'

29. The difficulty however is this: What can opening the window be except making such-and-such movements with such-and-such a result? And in that case what can *knowing* one is opening the window be except knowing that that is taking place? Now if there are two *ways* of knowing here, one of which I call knowledge of one's intentional action and the other of which I call knowledge by observation of what takes place, then must there not be two *objects* of knowledge? How can one speak of two different knowledges of *exactly* the same thing? It is not that there are two descriptions of the same thing, both of which are known, as when one knows that something is red and that it is coloured; no, here the description, opening the window, is identical, whether it is known by observation or by its being one's intentional action.

I think that it is the difficulty of this question that has led some people to say that what one knows as intentional action is only the intention, or possibly also the bodily movement; and that the rest is known by observation to be the *result*, which was also willed in the intention. But that is a mad account; for the only sense I can give to ' willing ' is that in which I might stare at something and will it to move. People sometimes say that one can get one's arm to move by an act of will but not a matchbox; but if they mean ' Will a matchbox to move and it won't ', the answer is ' If I will my arm to move in that way, it won't ', and if they mean ' I can move my arm but not the matchbox ' the answer is that I can move the matchbox—nothing easier.

Another false avenue of escape is to say that I really ' do ' in the intentional sense whatever I think I am doing. E.g. if I think I am moving my toe, but it is not actually moving, then I am ' moving my toe ' in a certain sense, and as for what *happens*, of course I haven't any control over that except in an accidental sense. The essential thing is just what has gone on in me, and if what happens coincides with what I ' do ' in the sphere of intentions, that is just a grace of fate. This I think was Wittgenstein's thought in the *Tractatus* when he wrote: ' The world is independent of my will ' and

> ' Even if what we wish were always to happen, this would only be a grace of fate, for it is not any logical connexion between will and the world that would guarantee this, and as for the presumed physical connexion, we cannot will *that*.' (6.373, 6.374).

That is to say: assuming it not to exist, willing it will be ineffectual. And I think that this reasoning applies to the effectiveness of *any* act of will. Hence Wittgenstein wrote in his notebooks at this time: ' I am completely powerless '.

But this is nonsense too. For if nothing guarantees that the window gets opened when I ' opened the window ', equally nothing guarantees that my toe moves when I ' move my toe '; so the only thing that does happen is my intention; but where is that to be found? I mean: what is its vehicle? Is it formulated in words? And if so, what guarantees that I do form the words that I intend? for the formulation of the words is itself an intentional act. And if the intention has no vehicle that is guaranteed, then what is there left for it to be but a bombination in a vacuum?

I myself formerly, in considering these problems, came out with the formula: I *do* what *happens*. That is to say, when the description of what happens is the very thing which I should say I was doing, then there is no distinction between my doing and the thing's happening. But everyone who heard this formula found it extremely paradoxical and obscure. And I think the reason is this: what happens must be given by observation; but I have argued that my knowledge of what I do is not by observation. A very clear and interesting case of this is that in which I shut my eyes and write something. I can say what I am writing. And what I say I am writing will almost always in fact appear on the paper. Now here it is clear that my capacity to say what is written is not derived from any observation. In practice of course what I write will very likely not go on being very legible if I don't use my eyes; but isn't the role of all our observation-knowledge in knowing what we are doing like the role of the eyes in producing successful writing? That is to say, once given that we have knowledge or opinion about the matter in which we perform intentional actions, our observation is merely an aid, as the eyes are an aid in writing. Someone without eyes may go on writing with a pen that has no more ink in it; or may not realise he is going over the edge of the paper on to the table or overwriting lines already written; here is where the eyes are useful; but the essential thing he does, namely to write such-and-such, is done without the eyes. So without the eyes he knows what he writes; but the eyes help to assure him that what he writes actually gets legibly written. In face of this how can I say: I *do* what *happens*? If there are two ways of knowing there must be two different things known.

30. Before I make an end of raising difficulties, I will produce an example which shews that it is an error to try to push what is known by being the content of intention back and back; first to the bodily movement, then perhaps to the contraction of the muscles, then to the attempt to do the thing, which comes right at the beginning. The only description that I clearly know of what I am doing may be of something that is at a distance from me. It is not the case that I clearly know the movements I make, and the intention is just a result which I calculate and hope will follow on these movements.

Someone might express the view I reject by saying: Consider

the sentence ' I am pushing the boat out '. Here, the only part of the sentence which really expresses the known action *in* this intentional action is ' I am pushing '. The words ' the boat ' express an opinion on an object which I take to be just in front of me; and that is verified by the senses, i.e. it is a matter of observation. The word ' out ' expresses intention with which I am pushing because it expresses an opinion as to an effect of my pushing in these circumstances, which opinion is accompanied by a desire on my part. And this must be the model for analysing every description of an intentional action.

My example to refute such a view is this. Imagine raising the following rather curious question: Is there any difference between letting one's arm drop and lowering one's arm at the speed at which it would fall? Can I deliberately lower my arm at the speed at which it would fall? I should find it difficult to make that the title under which I acted. But suppose someone simply wanted to produce the effect that in fact I lowered my arm at the speed at which it would fall—he is a physiologist, and wants to see if I generate anything different in the nerve fibres if I do this. So he fixes up a mechanism in which something in motion can be kept level if I hold a handle and execute a pumping movement with my arm and on the downward stroke lower it at the rate at which it would fall. Now my instruction is: Keep it level, and with a bit of practice I learn to do so. My account of what I am doing is that I am keeping the thing level; I don't consider the movement of my arm at all. I am able to give a much more exact account of what I am doing at a distance than of what my arm is doing. So my keeping the thing level is not at all something which I calculate as the effect of what I really and immediately am doing, and therefore directly know in my ' knowledge of my own action '. In general, as Aristotle says, one does not deliberate about an acquired skill; the description of what one is doing, which one completely understands, is at a distance from the details of one's movements, which one does not consider at all.

31. Having raised enough difficulties, let us try to sketch a solution, and let us first ask: What is the contradictory of a

description of one's own intentional action? Is it ' You aren't, in fact '?—E.g. ' You aren't replenishing the house water supply, because the water is running out of a hole in the pipe '? I suggest that it is not. To see this, consider the following story, which appeared for the pleasure of readers of the *New Statesman*'s ' This England ' column. A certain soldier was court-martialled (or something of the sort) for insubordinate behaviour. He had, it seems, been ' abusive ' at his medical examination. The examining doctor had told him to clench his teeth; whereupon he took them out, handed them to the doctor and said ' You clench them '.

Now the statement: ' The water is running out of a pipe round the corner ' stands in the same relation to the statement ' I'm replenishing the house water-supply ' as does ' My teeth are false ' to the order ' Clench your teeth '; and so the statement (on grounds of observation) ' You are not replenishing the house water-supply ' stands in the same relation to the description of intentional action ' I am replenishing the house water-supply ', as does the well-founded prediction ' This man isn't going to clench his teeth, since they are false ' to the order ' Clench your teeth '. And just as the contradiction of the order: ' Clench your teeth ' is *not* ' The man, as is clear from the following evidence, is not going to do any clenching of teeth, at least of the sort you mean ', but ' Do not clench your teeth ', so the contradiction of ' I'm replenishing the house water-supply ' is not ' You aren't, since there is a hole in the pipe ', but ' Oh, no, you aren't ' said by someone who thereupon sets out e.g. to make a hole in the pipe with a pick-axe. And similarly, if a person says ' I am going to bed at midnight ' the contradiction of this is not: ' You won't, for you never keep such resolutions ' but ' You won't, for I am going to stop you '.

But, returning to the order and the description by the agent of his present intentional action, is there not a point at which the parallelism ceases: namely, just where we begin to speak of knowledge? For we say that the agent's description is a piece of knowledge, but an order is not a piece of knowledge. So though the parallelism is interesting and illuminates the periphery of the problem, it fails at the centre and leaves that in the darkness that we have found ourselves in.

32. Let us consider a man going round a town with a shopping list in his hand. Now it is clear that the relation of this list to the things he actually buys is one and the same whether his wife gave him the list or it is his own list; and that there is a different relation when a list is made by a detective following him about. If he made the list itself, it was an expression of intention; if his wife gave it him, it has the role of an order. What then is the identical relation to what happens, in the order and the intention, which is not shared by the record? It is precisely this: if the list and the things that the man actually buys do not agree, and if this and this alone constitutes a *mistake*, then the mistake is not in the list but in the man's performance (if his wife were to say: ' Look, it says butter and you have bought margarine ', he would hardly reply: ' What a mistake! we must put that right ' and alter the word on the list to ' margarine '); whereas if the detective's record and what the man actually buys do not agree, then the mistake is in the record.

In the case of a discrepancy between the shopping list and what the man buys, I have to introduce the qualification: If this and this alone constitutes a mistake. For the discrepancy might arise because some of the things were not to be had and if one might have known they were not to be had, we might speak of a mistake (an error of judgment) in constructing the list. If I go out in Oxford with a shopping list including ' tackle for catching sharks ', no one will think of it as a mistake in performance that I fail to come back with it. And then again there may be a discrepancy between the list and what the man bought because he changed his mind and decided to buy something else instead.

This last discrepancy of course only arises when the description is of a future action. The case that we now want to consider is that of an agent who says what he is at present doing. Now suppose what he says is not true. It may be untrue because, unknown to the agent, something is not the case which would have to be the case in order for his statement to be true; as when, unknown to the man pumping, there was a hole in the pipe round the corner. But as I said, this relates to his statement that he is replenishing the water-supply as does the fact that the man has no teeth of his own to the order ' Clench your teeth ' ;

that is, we may say that in face of it his statement falls to the ground, as in that case the order falls to the ground, but it is not a direct contradiction. But is there not possible another case in which a man is *simply* not doing what he says? As when I say to myself ' Now I press Button A'—pressing Button B—a thing which can certainly happen. This I will call the *direct* falsification of what I say. And here, to use Theophrastus' expression again, the mistake is not one of judgment but of performance. That is, we do *not* say: What you *said* was a mistake, because it was supposed to describe what you did and did not describe it, but: What you *did* was a mistake, because it was not in accordance with what you said.

It is precisely analogous to obeying an order wrong—and we ought to be struck by the fact that there is such a thing, and that it is not the same as ignoring, disregarding, or disobeying an order. If the order is given ' Left turn! ' and the man turns right, there can be clear signs that this was not an act of disobedience. But there is a discrepancy between the language and that of which the language is a description. But the discrepancy does not impute a fault to the language—but to the event.

Can it be that there is something that modern philosophy has blankly misunderstood: namely what ancient and medieval philosophers meant by *practical knowledge*? Certainly in modern philosophy we have an incorrigibly contemplative conception of knowledge. Knowledge must be something that is judged as such by being in accordance with the facts. The facts, reality, are prior, and dictate what is to be said, if it is knowledge. And this is the explanation of the utter darkness in which we found ourselves. For if there are two knowledges—one by observation, the other in intention—then it looks as if there must be two objects of knowledge; but if one says the objects are the same, one looks hopelessly for the different *mode of contemplative knowledge* in acting, as if there were a very queer and special sort of seeing eye in the middle of the acting.

33. The notion of ' practical knowledge ' can only be understood if we first understand ' practical reasoning '. ' Practical reasoning ', or ' practical syllogism ', which means the same

thing, was one of Aristotle's best discoveries. But its true character has been obscured. It is commonly supposed to be ordinary reasoning leading to such a conclusion as: ' I ought to do such-and-such.' By ' ordinary reasoning ' I mean the only reasoning ordinarily considered in philosophy: reasoning towards the truth of a proposition, which is supposedly shewn to be true by the premises. Thus: ' Everyone with money ought to give to a beggar who asks him; this man asking me for money is a beggar; I have money; so I ought to give this man some '. Here the conclusion is entailed by the premises. So it is proved by them, unless they are doubtful. Perhaps such premises never can be certain.

Contemplating the accounts given by modern commentators, one might easily wonder why no one has ever pointed out the mince pie syllogism: the peculiarity of this would be that it was about mince pies, and an example would be 'All mince pies have suet in them—this is a mince pie—therefore etc.' Certainly ethics is of importance to human beings in a way that mince pies are not; but such importance cannot justify us in speaking of a special sort of reasoning. Everyone takes the practical syllogism to be a proof—granted the premises and saving their inevitable uncertainty or doubtfulness in application—of a conclusion. This is so whether Aristotle's own example has been taken:

> Dry food suits any human
> Such-and-such food is dry
> I am human
> This is a bit of such-and-such food

yielding the conclusion

> This food suits me

or whether, adopting a style of treatment suggested by some modern authors, the first premise is given in an imperative form. We may note that authors always use the term ' major ' and ' minor ' of the premises of practical syllogism: having regard to the definition of these terms, we can see that they have no application to Aristotle's practical syllogism, though they could be adapted to the imperative form if we assimilate ' Do ! ' to the predicate of a proposition. Consider the following:

Do everything conducive to not having a car crash.
Such-and-such will be conducive to not having a car crash.
Ergo: Do such-and-such.

Both this and the Aristotelian example given before would necessitate the conclusion. Someone professing to accept the opening order and the factual premise in the imperative example must accept its conclusion, just as someone believing the premises in the categorical example must accept its conclusion. The first example has the advantage of actually being Aristotle's, apart from the conclusion, but the disadvantage, so far as its being practical is concerned, that though the conclusion is necessitated, nothing seems to follow about doing anything. Many authors have pointed this out, but have usually put it rather vaguely, saying, e.g. that the reasoning does not compel any action; but Aristotle appears to envisage an action as following. The vague accounts that I have mentioned can be given a quite sharp sense. It is obvious that I can decide, on general grounds about colouring and so on, that a certain dress in a shop window would suit me very well, without its following that I can be accused of some kind of inconsistency with what I have decided if I do not thereupon go in and buy it; even if there are no impediments, such as shortage of cash, at all. The syllogism in the imperative form avoids this disadvantage; someone professing to accept the premises will be inconsistent if, when nothing intervenes to prevent him, he fails to act on the particular order with which the argument ends. But this syllogism suffers from the disadvantage that the first, universal, premise is an insane one,[1] which no one could accept for a moment if he thought out what it meant. For there are usually a hundred different and incompatible things conducive to not having a car crash; such as, perhaps, driving into the private gateway immediately on your left and abandoning your car there, and driving into the private gateway immediately on your right and abandoning the car there.

 The cause of this mischief, though it is not entirely his fault, is Aristotle himself. For he himself distinguished reasoning by subject matter as scientific and practical. ' Demonstrative ' reasoning was scientific and concerned what is invariable. As if one could not reason about some particular non-necessary thing that was going to happen except with a view to action! ' John

---

[1] No author, of course, has proposed this syllogism. I am indebted for the idea of it, however, to a passage in Mr. R. M. Hare's book, *The Language of Morals*, p. 35.

will drive from Chartres to Paris at an average of sixty m.p.h., he starts around five, Paris is sixty miles from Chartres, therefore he will arrive at about six '—this will not be what Aristotle calls a ' demonstration ' because, if we ask the question what John will do, that is certainly capable of turning out one way or another. But for all that the reasoning is an argument that something is true. It is not practical reasoning: it has not the form of a calculation what to do, though like any other piece of ' theoretical ' argument it could play a part in such a calculation. Thus we may accept from Aristotle that practical reasoning is essentially concerned with ' what is capable of turning out variously ', without thinking that this subject matter is enough to make reasoning about it practical. There is a difference of form between reasoning leading to action and reasoning for the truth of a conclusion. Aristotle however liked to stress the similarity between the kinds of reasoning, saying[1] that what ' happens ' is the same in both. There are indeed three types of case. There is the theoretical syllogism and also the idle practical syllogism[2] which is just a classroom example. In both of these the conclusion is ' said ' by the mind which infers it. And there is the practical syllogism proper. Here the conclusion is an action whose point is shewn by the premises, which are now, so to speak, on active service. When Aristotle says that what happens is the same, he seems to mean that it is always the same psychical mechanism by which a conclusion is elicited. He also displays practical syllogisms so as to make them look as parallel as possible to proof syllogisms.

Let us imitate one of his classroom examples, giving it a plausible modern content:

Vitamin X is good for all men over 60
Pigs' tripes are full of vitamin X
I'm a man over 60
Here's some pigs' tripes.

Aristotle seldom states the conclusion of a practical syllogism, and sometimes speaks of it as an action; so we may suppose the man who has been thinking on these lines to take some of the dish that he sees. But there is of course no objection to inventing a

[1] *De Motu Animalium VII.*   [2] *Ethica Nicomachea* 1147a, 27-8.

form of words by which he *accompanies* this action, which we may
call the conclusion in a verbalised form. We may render it as:

(*a*) So I'll have some
or (*b*) So I'd better have some.
or (*c*) So it'd be a good thing for me to have some.

Now certainly no one could be tempted to think of (*a*) as a
proposition entailed by the premises. But neither are (*b*) and (*c*),
though at first sight they look roughly similar to the kind of
conclusion which commentators usually give:

What's here is good for me.

But of course in the sense in which this is entailed by the premises
as they intend it to be, this only means: ' What's here is a type of
food that is good for me ', which is far from meaning that I'd
better have some. Now the reason why we cannot extract ' I'd
better have some ' from the premises is not at all that we *could not*
in any case construct premises which, if assented to, yield this
conclusion. For we could, easily. We only need to alter the
universal premise slightly, to:

It is necessary for all men over 60 to eat any food containing Vitamin X that they ever come across

which, with the other premises, would entail the conclusion in
the form ' I'd better have some ' quite satisfactorily. The only
objection is that the premise is insane, as would have been the
corresponding variant on Aristotle's universal premise:

Every human being needs to eat all the dry food he
ever sees.

In short the ' universality ' of Aristotle's universal premise is
in the wrong place to yield the conclusion by way of entailment
at all.

Only negative general premises can hope to avoid insanity
of this sort. Now these, even if accepted as practical premises,
don't lead to any particular actions (at least, not by themselves
or by any formal process) but only to not doing certain things.
But what Aristotle meant by practical reasoning certainly in-

cluded reasoning that led to action, not to omissions. Now a man who goes through such considerations as those about Vitamin X and ends up by taking some of the dish that he sees, saying e.g. ' So I suppose I'd better have some ', can certainly be said to be *reasoning*; on the other hand, it is clear that this is another type of reasoning than reasoning from premises to a conclusion which they prove. And I think it is even safe to say that (except in, say, doing arithmetic or dancing, i.e. in skills or arts—what Aristotle would call τέχναι) there is no general positive rule of the form 'Always do X ' or ' Doing X is always good—required—con-venient—, a useful—suitable—etc.—thing ' (where the ' X ' describes some specific action) which a sane person will accept as a starting-point for reasoning out what to do in a particular case. (Unless, indeed, it is hemmed about by saving clauses like ' if the circumstances don't include something that would make it foolish '.) Thus though general considerations, like ' Vitamin C is good for people ' (which of course is a matter of medical fact) may easily occur to someone who is considering what he is going to eat, considerations of the form ' Doing such-and-such quite specific things in such-and-such circumstances is always suitable ' are never, if taken strictly, possible at all for a sane person, outside special arts.

34. But, we may ask, even if we want to follow Aristotle, need we confine the term ' practical reasoning ' to pieces of practical reasoning which look very parallel to proof-reason-ings? For ' I want a Jersey cow; there are good ones in the Hereford market, so I'll go there ' would seem to be practical reasoning too. Or ' If I invite both X and Y, there'll be a strained atmosphere in view of what X has recently said about Y and how Y feels about it—so I'll just ask X '. Or again ' So-and-so was very pleasant last time we met, so I'll pay him a visit '. Now Aristotle would have remarked that it is mere ' desire ' in a special sense (ἐπιθυμία) that prompts the action in the last case; the mark of this is that the premise refers to something merely as pleasant. The point that he is making here is, however, rather alien to us, since we do not make much distinction between one sort of desire and another, and we should say: isn't it desire in some sense—i.e. wanting—that prompts the action in all the cases?

And ' all cases ', of course, includes ones that have as large an apparatus as one pleases of generalisations about morals, or medicine, or cookery, or methods of study, or methods of getting votes or securing law and order, together with the identification of cases.

This is so, of course, and is a point insisted on by Aristotle himself: the ἀρχή (starting point) is τὸ ὀρεκτόν (the thing wanted). For example, the fact that current school geometry text books all give a faulty proof of the theorem about the base angles of an isosceles triangle will not lead a teacher to discard them or to make a point of disabusing his class, if he does not want to impart *only* correct geometrical proofs. He will say that it doesn't matter; the Euclidean proof, Pons Asinorum, is too difficult; in any case Euclid starts (he may say) with the unjustified assumption that a certain pair of circles will cut; and are you going to suggest worries about the axiom of parallels to school children and try to teach them non-Euclidean geometry? and much else of the sort. All this obscures the essential point, which is that, rightly or wrongly, he does not want to impart *only* correct geometrical reasoning. It then becomes relevant to ask what he does want to do. Let us suppose that he is reasonably frank and says he wants to keep his job, occupy his time in ' teaching ', and earn his salary.

This question ' What do you want? ' was not a question out of the blue, like ' What are the things you want in life? ' asked in a general way at the fireside. In context, it is the question ' With a view to what are you doing X, Y and Z? ', which are what he is doing. That is to say, it is a form of our question ' Why? ' but with a slightly altered appearance. If a man is asked *this* question about what he is doing, that ' with a view to which ' he does it is always beyond the break at which we stopped in §23. For even if a man ' is doing ' what he ' wants ', like our imaginary teacher, he has never completely attained it, unless by the termination of the time for which he wants it (which might be the term of his life).

35. In four practical syllogisms that Aristotle gives us, there occur the expressions ' it suits ', ' should ', and ' pleasant '. The four universal premises in question are:

(*a*) Dry food suits any man
(*b*) [I] should taste everything sweet
(*c*) Anything sweet is pleasant
(*d*) Such a one should do such a thing

The first three come from the *Nicomachean Ethics*, the fourth from the *De Anima*; in the *De Anima* Aristotle is discussing what sets a human being in physical motion, and this universal (*d*) is just a schema of a universal premise. The occurrence of ' should ' in it has no doubt helped the view that the practical syllogism is essentially ethical, but the view has no plausibility; this is not an ethical passage, and Aristotle nowhere suggests that the starting point is anything but something *wanted*. In thinking of the word for ' should ' ' ought ' etc. (δεῖ) as it occurs in Aristotle, we should think of it as it occurs in ordinary language (e.g. as it has just occurred in this sentence) and not just as it occurs in the examples of ' moral discourse ' given by moral philosophers. That athletes should keep in training, pregnant women watch their weight, film stars their publicity, that one should brush one's teeth, that one should (not) be fastidious about one's pleasures, that one should (not) tell ' necessary ' lies, that chairmen in discussions should tactfully suppress irrelevancies, that someone learning arithmetic should practise a certain neatness, that machinery needs lubrication, that meals ought to be punctual, that we should (not) see the methods of ' Linguistic Analysis' in Aristotle's philosophy; any fair selection of examples, if we care to summon them up, should convince us that ' should ' is a rather light word with unlimited contexts of application, and it can be presumed that it is because of this feature that Aristotle chose a roughly corresponding Greek word as the word to put into the universal premise of his schematic practical syllogism. Case (*b*) appears to presuppose a situation where one is given this premise—it is, say, an instruction to an undercook in a kitchen in a special eventuality. Aristotle is here[1] giving us a futile mechanistic theory of how premises work to produce a conclusion: e.g. given this curious premise and the information ' this is sweet ' together, the action of tasting it is mechanically produced if there is nothing to stop it. We notice that this

---

[1] *Ethica Nicomachea* 1147a 28.

premise has the universality required to necessitate the conclusion for someone who accepts it; just for that reason it is absurd unless restricted to a particular situation—or unless we are to imagine someone having a sweet tooth to the point of mania.

Thus there is nothing necessarily ethical about the word ' should ' occurring in the universal premise of a practical syllogism, at least so far as concerns the remarks made by Aristotle who invented the notion. But we find ' should ' ' suits ' or ' pleasant ' (or some other evaluative term) in all the examples that he gives, and it is reasonable to ask why. If the starting point for a practical syllogism is something wanted, then why should the first premise not be ' I want . . . ' as in the example ' I want a Jersey cow '? The case as I imagined it is surely one of practical reasoning.

But it is misleading to put ' I want ' into a premise if we are giving a formal account of practical reasoning. To understand this, we need to realise that not everything that I have described as coming in the range of ' reasons for acting ' can have a place as a premise in a practical syllogism. E.g. ' He killed my father, so I shall kill him ' is not a form of reasoning at all; nor is ' I admire him so much, I shall sign the petition he is sponsoring '. The difference is that there is no calculation in these. The conjunction ' so ' is not necessarily a mark of calculation.

It may be said: ' if " he was very pleasant . . . so I shall pay him a visit " can be called reasoning, why not " I admire . . . so I shall sign "? '. The answer is that the former is not a piece of reasoning or calculation either, if what it suggests is e.g. that I am making a return for his pleasantness, have this reason for the kind act of paying a visit; but if the suggestion is: ' So it will probably be pleasant to see him again, so I shall pay him a visit ', then it is; and of course it is only under this aspect that ' desire ' in the restricted sense (ἐπιθυμία) is said to prompt the action. And similarly: ' I admire . . . and the best way to express this will be to sign, so I shall sign . . . ' is a case of calculating, and if that is the thought we can once again speak of practical reasoning. Of course ' he was pleasant . . . How can I make a return? . . . I will visit him ' can occur and so this case assume the form of a calculation. Here a return, *under that description*, becomes the

F

object of wish; but what is the meaning of 'a return'? The primitive, spontaneous, form lies behind the formation of the concept 'return', which *once formed* can be made the object of wish; but in the primitive, spontaneous, case the form is 'he was nice to me—I will visit him'; and similarly with revenge, though once the concept 'revenge' exists it can be made the object, as with Hamlet. We must always remember that an object is not what what is aimed at *is*; the description *under which* it is aimed at is that under which it is *called* the object.

Then 'I want this, so I'll do it' is not a form of practical reasoning either. The role of 'wanting' in the practical syllogism is quite different from that of a premise. It is that whatever is described in the proposition that is the starting-point of the argument must be wanted in order for the reasoning to lead to any action. Then the form 'I want a Jersey cow, they have good ones in the Hereford market, so I'll go there' was formally misconceived: the practical reasoning should just be given in the form 'They have Jersey cows in the Hereford market, so I'll go there'. Similarly 'Dry food' (whatever Aristotle meant by that; it sounds an odd dietary theory) 'suits anyone etc., so I'll have some of this' is a piece of reasoning which will go on only in someone who wants to eat suitable food. That is to say, it will at any rate terminate in the conclusion only for someone who wants to eat suitable food. Someone free of any such wish might indeed calculate or reason up to the conclusion, but leave that out, or change it to—'So eating this would be a good idea (if I wanted to eat suitable food).' Roughly speaking we can say that the reasoning leading up to an action would enable us to infer what the man so reasoning wanted—e.g. that he probably wanted to see, buy, or steal a Jersey cow.

There is a contrast between the two propositions 'They have some good Jerseys in the Hereford market' and 'Dry food suits any man', supposing that they both occur as practical premises, i.e. that the man who uses the one sets off for Hereford, and the man who uses the other takes a bit of the dish that he sees, believing it to be a bit of some kind of dry food. In the first case, there can arise the question 'What do you want a Jersey cow for?'; but the question 'What do you want suitable food for?' means, if anything 'Do give up thinking about food as suitable or otherwise '—as said e.g. by someone who prefers people merely to enjoy their food or considers the man hypochondriac.

36. It is a familiar doctrine that people can want anything; that is, that in 'A wants X' 'X' ranges over all describable objects or states of affairs. This is untenable; for example the range is restricted to present or future objects and future states of affairs; for we are not here concerned with idle wishing. A chief mark of an idle wish is that a man does nothing—whether he could or no—towards the fulfilment of the wish. Perhaps the familiar doctrine I have mentioned can be made correct by being restricted to wishing. The most primitive expression of wishing is e.g. 'Ah, if only . . .!'—if only $\sqrt{2}$ were commensurable, or Helen were still alive, or the sun would blow up, or I could hold the moon in the palm of my hand, or Troy had not fallen, or I were a millionaire. It is a special form of expression, to which a characteristic tone of voice is appropriate; and it might be instructive to ask how such a form is identified (e.g. in a language learnt in use); but it does not concern us here.

' Wanting ' may of course be applied to the prick of desire at the thought or sight of an object, even though a man then does nothing towards getting the object. Now where an object which arouses some feeling of longing is some future state of affairs of which there is at least some prospect, wanting, as the longing may be called if it is sustained, may be barely distinguished from idle wishing; the more the thing is envisaged as a likelihood, the more wishing turns into wanting—if it does not evaporate at the possibility. Such wanting is hope. But wanting, in the sense of the prick of desire, is compatible with one's doing nothing at all towards getting what one wants, even though one could do something; while to hope that something will happen that it is in one's power to try to bring about, and yet do nothing to bring it about, is hope of a rather degenerate kind; or ' *hope* that it will happen', though I do none of the things I know I might do towards it, is rather ' hope that it will happen *without* my doing anything towards it ': a different object from that of the first hope.

The wanting that interests us, however, is neither wishing nor hoping nor the feeling of desire, and cannot be said to exist in a man who does nothing towards getting what he wants.

The primitive sign of wanting is *trying to get*; which of course can only be ascribed to creatures endowed with sensation. Thus it is not mere movement or stretching out towards something, but this on the part of a creature that can be said to know the thing. On the other hand knowledge itself cannot be described independently of volition; the ascription of sensible knowledge and of volition go together. One idea implicit in phenomenalism has always been that e.g. the knowledge of the meaning of colour-words is only a matter of picking out and naming certain perceived differences and similarities between objects. And this kind of idea is not dead even though phenomenalism is not fashionable. A modern Psammetichus, influenced by epistemologists, might have a child cared for by people whose instructions were to make no sign to the child in *dealing* with it, but frequently to utter the names of the objects and properties which they judged to be within its perceptual fields, with a view to finding out which were the very first things or properties that humans learned to name. But e.g. the identification served by colour-names is in fact not primarily that of colours, but of objects by means of colours; and thus, too, the prime mark of colour-discrimination is doing things with objects—fetching them, carrying them, placing them—according to their colours. Thus the possession of sensible discrimination and that of volition are inseparable; one cannot describe a creature as having the power of sensation without also describing it as doing things in accordance with perceived sensible differences. (Naturally this does not mean that every perception must be accompanied by some action; it is because that is not so that it is possible to form an epistemology according to which the names of the objects of perception are just given in some kind of ostensive definition.)

The primitive sign of wanting is *trying to get*: in saying this, we describe the movement of an animal in terms that reach beyond what the animal is now doing. When a dog smells a piece of meat that lies the other side of the door, his trying to get it will be his scratching violently round the edges of the door and snuffling along the bottom of it and so on. Thus there are two features present in wanting; movement towards a thing and knowledge (or at least opinion) that the thing is there.

When we consider human action, though it is a great deal more complicated, the same features are present when what is wanted is something that already exists: such as a particular Jersey cow, which is presumed to be on sale in the Hereford market, or a particular woman desired in marriage.

But a man can want *a* cow, not any particular cow, or *a* wife. This raises a difficulty best expressed from the point of view of the theory of descriptions. For we cannot render 'A wants a cow' as 'It is not always false of x that x is a cow and A wants x'. Nor can we get out of this difficulty by introducing belief into our analysis and then using what Russell says about belief: namely that 'A believes that a cow is in the garden' can mean, not, 'It is not always false of x that x is a cow and A believes that x is in the garden' but 'A believes that it is not always false of x . . .' For, plainly, wanting a cow need not involve a belief 'some cow is—'; and still less does wanting a wife involve a belief 'some wife of mine is—'. A similar difficulty can indeed arise for animals too: we say the cat is waiting for *a* mouse at a mousehole, but suppose there is *no* mouse? Here, however, it *is* reasonable enough to introduce belief and say that the cat *thinks* there is a mouse: I intend such an expression just as it would quite naturally be said. And though it seems rather comical to apply Russell's analysis to the 'thoughts' of a cat, there is not really any objection; for our difficulty was a logical one, about the status of the denoting phrase 'a mouse' in 'the cat is waiting for a mouse', and not one about what may go on in the souls of cats; hence Russell's analysis can be used to dispel the difficulty. And when we say 'The dog wants a bone' there is not much difficulty either; for we can say that the dog knows that there are bones in a bag and is excited and so on, or that he always gets a bone at this time and so is in a state of excitement and dissatisfaction until he gets one. But when a man wants a wife, there seems to be greater difficulty. We must say: he wants 'It is not always false of x . . .' to *become* true. (Here I depart from Russell in holding that propositions can be variable in truth-value; I should do that in any case, on other grounds. But in consequence the word 'always' becomes slightly misleading, and so I would substitute the commoner form: It is not for all x not the case that . . .)

Thus the special problems connected with indefinite descriptions do not turn out to create peculiar difficulties for an account of wanting; the difficulty here is the general one that arises when the object of wanting is not anything that exists or that the agent supposes to exist. For we spoke of two features present in ' wanting ' : movement towards something, and knowledge, or at least opinion, that the thing is there. But where the thing wanted is not even supposed to exist, as when it is a future state of affairs, we have to speak of an idea, rather than of knowledge or opinion. And our two features become: some kind of action or movement which (the agent at least supposes) is of use towards something, and the idea of that thing.

The other senses of ' wanting ' which we have noticed are not of any interest in a study of action and intention.

37.   Are there any further restrictions, besides the ones we have mentioned, on possible objects of wanting, when the idea of the thing that is (in fact) wanted is expressed in the first premise of a practical syllogism? There are, we may say, no further absolute restrictions, but there are some relative ones. For, as I have remarked, if ' There are good Jerseys in the Hereford market ' is used as a premise, then it can be asked ' What do you want a Jersey for? '. Let the answer be: 'A Jersey would suit my needs well '.—And it is in fact this or a form of this, that Aristotle would accept as first premise: the reasoning in his chosen form would run: '(1) Any farmer with a farm like mine could do with a cow of such-and-such qualities (2) e.g. a Jersey.' Now there is no room for a *further* question " What do you want ' what you could do with ' for? " That is to say, the premise now given has characterised the thing wanted as desirable.

But is not anything wantable, or at least any perhaps attainable thing? It will be instructive to anyone who thinks this to approach someone and say: ' I want a saucer of mud ' or ' I want a twig of mountain ash '. He is likely to be asked what for; to which let him reply that he does not want it *for* anything, he just wants it. It is likely that the other will then perceive that a philosophical example is all that is in question, and will pursue the matter no further; but supposing that he did not realise this, and yet did not dismiss our man as a dull babbling loon, would he not try to find out in what aspect the object desired is desirable? Does it serve as a symbol? Is there something delightful about it? Does the man want to have something to call his own, and no more? Now if the reply is: 'Philosophers have taught that anything can be an object of desire; so there can be no need for me to characterise these objects as somehow desirable; it merely so happens that I want them', then this is fair nonsense.

But cannot a man *try to get* anything gettable? He can certainly go after objects that he sees, fetch them, and keep them near him; perhaps he then vigorously protects them from removal. But then, this is already beginning to make sense: these are his possessions, he wanted to own them; he may be idiotic, but his 'wanting' is recognisable as such. So *he* can say perhaps 'I want a saucer of mud'. Now saying 'I want' is often a way to be given something; so when out of the blue someone says 'I want a pin' and denies wanting it *for* anything, let us suppose we give it him and see what he does with it. He takes it, let us say, he smiles and says 'Thank you. My want is gratified'—but what does he do with the pin? If he puts it down and forgets about it, in what sense was it true to say that he wanted a pin? He used these words, the effect of which was that he was given one; but what reason have we to say he wanted a pin rather than: to see if we would take the trouble to give him one?

It is not a mere matter of what is usual in the way of wants and what is not. It is not at all clear what it meant to say: this man simply wanted a pin. Of course, if he is careful always to carry the pin in his hand thereafter, or at least for a time, we may perhaps say: it seems he really wanted that pin. Then perhaps, the answer to 'What do you want it for?' may be 'to carry it about with me', as a man may want a stick. But here again there is further characterisation: 'I don't feel comfortable without it; it is pleasant to have one' and so on. To say 'I *merely* want this' without any characterisation is to deprive the word of sense; if he insists on 'having' the thing, we want to know what 'having' amounts to.

Then Aristotle's terms: 'should', 'suits', 'pleasant' are characterisations of what they apply to as desirable. Such a characterisation has the consequence that no further questions 'what for?', relating to the characteristic so occurring in a premise, require any answer. We have seen that at least sometimes a description of an object wanted is subject to such a question, i.e. such a question about the description does require an answer. This, then will be why Aristotle's forms of the practical syllogism give us such first premises.

Aristotle gives us a further practical syllogism when he remarks ' a man may know that light meats are digestible and wholesome but not know which meats are light '.[1] Here the description ' digestible and wholesome ' might seem not to be a pure desirability-characterisation. But since wholesome means good for the health, and health is by definition the *good* general state of the physical organism, the characterisation is adequate for a proper first premise and does not need to be eked out by, say, ' health is a human good ' (a tautology).

38. Let us now consider an actual case where a desirability characterisation gives a final answer to the series of ' What for? ' questions that arise about an action. In the present state of philosophy, it seems necessary to choose an example which is not obscured by the fact that moral approbation on the part of the writer or reader is called into play; for such approbation is in fact irrelevant to the logical features of practical reasoning; but if it is evoked, it may seem to play a significant part. The Nazis, being pretty well universally execrated, seem to provide us with suitable material. Let us suppose some Nazis caught in a trap in which they are sure to be killed. They have a compound full of Jewish children near them. One of them selects a site and starts setting up a mortar. Why this site?—Any site with such-and-such characteristics will do, and this has them. Why set up the mortar?—It is the best way of killing off the Jewish children. Why kill off the Jewish children?—It befits a Nazi, if he must die, to spend his last hour exterminating Jews. (I am a Nazi, this is my last hour, here are some Jews.) Here we have arrived at a desirability characterisation which makes an end of the questions ' What for ? '

Aristotle would seem to have held that every action done by a rational agent was capable of having its grounds set forth up to a premise containing a desirability characterisation; and as we

[1] *Ethica Nicomachea*, 1141 b 18.

have seen, there is a reasonable ground for this view, wherever
there is a calculation of means to ends, or of ways of doing what
one wants to do. Of course ' fun ' is a desirability characterisation
too, or ' pleasant ': ' Such-and-such a kind of thing is pleasant '
is one of the possible first premises. But cannot pleasure be taken
in *anything*? It all seems to depend on how the agent feels about
it! ' But *can* it be taken in anything? Imagine saying ' I want a
pin ' and when asked why, saying ' For fun '; or ' Because of the
pleasure of it '. One would be asked to give an account making it
at least dimly plausible that there was a pleasure here. Hobbes[1]
believed, perhaps wrongly, that there could be no such thing as
pleasure in mere cruelty, simply in another's suffering; but he
was not *so* wrong as we are likely to think. He was wrong in
suggesting that cruelty had to have an end, but it does have to
have a point. To depict this pleasure, people evoke notions of
power, or perhaps of getting one's own back on the world, or
perhaps of sexual excitement. No one needs to surround the
pleasures of food and drink with such explanations.

Aristotle's specifications for the action of a rational agent do
not cover the case of ' I just did, for no particular reason '. But
where this answer is genuine, there is no calculation, and therefore no intermediate premises (like 'Any site with such-and-such
characteristics will be a suitable one for setting up my mortar ',
and ' This is the best way to kill off the children') about which to
press the question ' What for? '. So we may note, as we have
done, that this sort of action ' for no particular reason ' exists,
and that here of course there is no desirability characterisation,
but that does not shew that the demand for a desirability characterisation, wherever there is a purpose at all, is wrong.

With ' It befits a Nazi, if he must die, to spend his last hour
exterminating Jews' we have then reached a terminus in enquiring
into that particular order of reasons to which Aristotle gave the
name ' practical '. Or again: we have reached the prime starting
point and can look no further. (The question ' Why be a Nazi? '
is not a continuation of *this* series; it addresses itself to one of
the particular premises.) Any premise, if it really works as a
first premise in a bit of ' practical reasoning ', contains a description of something wanted; but with the intermediary premises,

[1] *Leviathan* Part I, Chap. VI.

the question ' What do you want that for? ' arises—until at last
we reach the desirability characterisation, about which ' What do
you want that for? ' does not arise, or if it is asked has not the
same point, as we saw in the ' suitable food ' example.

But in saying this, I do not at all mean to suggest that there
is no such thing as taking exception to, or arguing against, the
first premise, or its being made the first premise. Nor am I
thinking of moral dissent from it; I prefer to leave that out of
account. But there are other ways of taking exception to, or
dissenting from, it. The first is to hold the premise false; as a
dietician might hold false Aristotle's views on dry food. It does
indeed befit a Nazi to exterminate Jews, the objector may say,
but there is a Nazi sacrament of dying which is what really befits
a Nazi if he is going to die, and has time for it. Or again the
objector may deny that it befits a Nazi as such to exterminate
Jews at all. However, both these denials would be incorrect,
so we may pass quickly on to other forms of demurrer. All of
these admit the truth of the proposition, and all but one oppose
the desire of what it mentions, namely to do what befits a Nazi
in the hour of death. The one that does not oppose it says: ' Yes,
that befits a Nazi, but so equally does such-and-such: why not
do something falling under *this* description instead, namely. . . '
Another says: ' To be sure, but at this moment I lose all interest
in doing what befits a Nazi '. And yet another says ' While that
does indeed befit a Nazi, it is not quite necessary for him to do it.
Nazism does not always require a man to strain to the utmost, it
is not as inhuman as that: no, it is quite compatible with being a
good Nazi to give yourself over to soft and tender thoughts of
your home, your family, and your friends, to sing our songs and
to drink the healths of those we love '. If any of these considerations work on him, the particular practical syllogism of our
original Nazi fails, though not on account of any falsehood in the
premise, even according to him, nor on account of any fault in
his practical calculation.

39. A (formal) ethical argument against the Nazi might
perhaps oppose the notion of ' What a *man* ought to do '[1] to

---

[1] But is it not perfectly possible to say: 'At this moment I lose all interest in
doing what befits a man '? If Aristotle thought otherwise, he was surely wrong.
I suspect that he thought a man could not lack this interest except under the influ-
ence of inordinate passion or through ' boorishness ' (ἀγροικία), i.e. insensibility.

the Nazi's original premise; setting up a position from which it followed incidentally that it did not befit a man to be a Nazi since a man ought not to do what befits a Nazi. Of course it is merely academic to imagine this; if the man with the moral objection were clever he would adopt one of the three last mentioned methods of opposing the hero, of which the first one would very likely be the best. But the following (vague) question is often asked in one form or another: if desirability characterisations are required in the end for purposive action, then must not the ones which relate to human good as such (in contrast with the good of film stars or shopkeepers) be in some obscure way compulsive, if believed? So someone who gets these right *must* be good; or at least (logically) *must* take a course within a certain permitted range or be ashamed. Some such idea too lies at the back of the notion that the practical syllogism is ethical.

' Evil be thou my good ' is often thought to be senseless in some way. Now all that concerns us here is that ' What's the good of it? ' is something that can be asked until a desirability characterisation has been reached and made intelligible. If then the answer to this question at some stage is ' The good of it is that it's bad ', this need not be unintelligible; one can go on to say 'And what is the good of its being bad? ' to which the answer might be condemnation of good as impotent, slavish, and inglorious. Then the good of making evil my good is my intact liberty in the unsubmissiveness of my will. *Bonum est multiplex*: good is multiform, and all that is required for our concept of ' wanting ' is that a man should see what he wants under the aspect of some good. A collection of bits of bone three inches long, if it is a man's object, is something we want to hear the praise of before we can understand it as an object; it would be affectation to say ' One can want anything and I *happen* to want this ', and in fact a collector does not talk like that; no one talks like that except in irritation and to make an end of tedious questioning. But when a man aims at health or pleasure, then the enquiry ' What's the good of it? ' is not a sensible one. As for reasons against a man's making one of them his principal aim; and whether there are orders of human goods, e.g. whether some are greater than others, and whether if this is so a man need ever prefer the greater to the less[1], and on pain of what; this question would belong to ethics, if there is such a science. All that I am concerned to argue here is that the fact that *some* desirability characterisation is required does not have the least tendency to shew that *any* is endowed with some kind of necessity in relation to wanting. But it may still be true that the man who says ' Evil be thou my good ' in the way that we described is committing errors of thought; this question belongs to ethics.

40. The conceptual connexion between ' wanting ' (in the sense which we have isolated, for of course we are not speaking of the ' I want ' of a child who screams for something) and ' good' can be compared to the conceptual connexion between ' judgment ' and ' truth '. Truth is the object of judgment, and good the object of wanting; it does not follow from this either that everything judged must be true, or that everything wanted must be good. But there is a certain contrast between these pairs of concepts too. For you cannot explain truth without introducing as its subject intellect, or judgment, or propositions, in some relation of which to the things known or judged truth consists; ' truth ' is ascribed to what has the relation, not to the things. With ' good ' and ' wanting ' it is the other way round; as we have seen, an account of ' wanting ' introduces good as its object, and goodness of one sort or another is ascribed primarily to the objects, not to the wanting: one wants a *good kettle*, but has a *true idea* of a kettle (as opposed to wanting a kettle well, or having an idea of a true kettle). Goodness is ascribed to wanting in virtue of the goodness (not the actualisation) of what is wanted; whereas truth is ascribed immediately to judgments, and in virtue of what actually *is* the case. But again, the notion of ' good ' that has to be introduced in an account of wanting is not that of what is really good but of what the agent conceives to be good ; what the agent wants would have to be characterisable as good by him, if we may suppose him not to be impeded by inarticulateness. Whereas when we are explaining truth as a predicate of judgments, propositions, or thoughts, we have to speak of a relation to what is really so, not just of what seems so to the judging mind. But on the other hand again, the good

[1] Following Hume, though without his animus, I of course deny that this preference can be as such ' required by reason ', in any sense.

(perhaps falsely) conceived by the agent to characterise the thing must *really* be one of the many forms of good.

We have long been familiar with the difficulties surrounding a philosophical elucidation of judgment, propositions, and truth; but I believe that it has not been much noticed in modern philosophy that comparable problems exist in connexion with 'wanting' and 'good'. In consequence there has been a great deal of absurd philosophy both about this concept and about matters connected with it.

The cause of blindness to these problems seems to have been the epistemology characteristic of Locke, and also of Hume. Any sort of wanting would be an internal impression according to those philosophers. The bad effects of their epistemology come out most clearly if we consider the striking fact that the concept of pleasure has hardly seemed a problematic one at all to modern philosophers, until Ryle reintroduced it as a topic a year or two ago.[1] The ancients seem to have been baffled by it; its difficulty, astonishingly, reduced Aristotle to babble, since for good reasons he both wanted pleasure to be identical with and to be different from the activity that it is pleasure in. It is customary nowadays to refute utilitarianism by accusing it of the 'naturalistic fallacy', an accusation whose force I doubt. What ought to rule that philosophy out of consideration at once is the fact that it always proceeds as if 'pleasure' were a quite unproblematic concept. No doubt it was possible to have this assumption because the notion that pleasure was a particular internal impression was uncritically inherited from the British empiricists. But it shews surprising superficiality both to accept that notion and to treat pleasure as quite generally the point of doing anything. We might adapt a remark of Wittgenstein's about meaning and say 'Pleasure cannot be an impression; for no impression could have the consequences of pleasure'. They were saying that something which they thought of as like a particular tickle or itch was quite obviously the point of doing anything whatsoever.

In this enquiry I leave the concept 'pleasure' in its obscurity; it needs a whole enquiry to itself.[2] Nor should an unexamined

---

[1] Aristotelian Society Supplementary Volume XXVIII, 1954.
[2] Aristotle's use of an artificial concept of 'choice', where I use 'intention', in describing 'action', is linked with the difficulty of this topic.

thesis 'pleasure is good' (whatever that may mean) be ascribed to me. For my present purposes all that is required is that 'It's pleasant' is an adequate answer to 'What's the good of it?' or 'What do you want that for?' I.e., the chain of 'Why's' comes to an end with this answer. The fact that a claim *that* 'it's pleasant' can be challenged, or an explanation asked for ('But what *is* the pleasure of it?') is a different point, as also would be any consideration, belonging properly to ethics, of its decency as an answer.

41. It will have become clear that the practical syllogism as such is not an ethical topic. It will be of interest to an ethicist, perhaps, if he takes the rather unconvincing line that a good man is by definition just one who aims wisely at good ends. I call this unconvincing because human goodness suggests virtues among other things, and one does not think of choosing means to ends as obviously the whole of courage, temperance, honesty, and so on. So what can the practical syllogism have to do with ethics? It can only come into ethical studies if a correct philosophical psychology is requisite for a philosophical system of ethics: a view which I believe I should maintain if I thought of trying to construct such a system; but which I believe is not generally current. I am not saying that there cannot be any such thing as moral general premises, such as 'People have a duty of paying their employees promptly', or Huckleberry Finn's conviction, which he failed to make his premise: 'White boys ought to give runaway slaves up'; obviously there can, but it is clear that such general premises will only occur as premises of practical reasoning in people who want to do their duty.[1] The point is very obvious, but has been obscured by the conception of the practical syllogism as of its nature ethical, and thus as a proof about what one ought to do, which somehow naturally culminates in action.

---

[1] It is worth remarking that the concepts of 'duty' and 'obligation', and what is now called the 'moral' sense of 'ought', are survivals from a *law* conception of ethics. The modern sense of 'moral' is itself a late derivative from these survivals. None of these notions occur in Aristotle. The idea that actions which are necessary if one is to conform to justice and the other virtues are requirements of divine law was found among the Stoics, and became generally current through Christianity, whose ethical notions come from the Torah.

Of course ' I ought to do this, so I'll do it ' is not a piece of practical reasoning any more than ' This is nice, so I'll have some ' is. The mark of practical reasoning is that the thing wanted is *at a distance* from the immediate action, and the immediate action is calculated as the way of getting or doing or securing the thing wanted. Now it may be at a distance in various ways. For example, ' resting ' is merely a wider description of what I am perhaps doing in lying on my bed; and acts done to fulfil moral laws will generally be related to positive precepts in this way; whereas getting in the good government is remote in time from the act of pumping, and the replenishment of the house water-supply, while very little distant in time, is at some spatial distance from the act of pumping.

42. We have so far considered only a particular unit of practical reasoning, to which the expression ' practical syllogism ' is usually restricted. But of course ' practical syllogisms ' in Greek simply means practical reasonings, and these include reasonings running from an objective through many steps to the performance of a particular action here and now. E.g. an Aristotelian doctor wants to reduce a swelling; this he says will be done by producing a certain condition of the blood; this can be produced by applying a certain kind of remedy; such-and-such a medicine is that kind of remedy; here is some of that medicine—give it.

It has an absurd appearance when practical reasonings, and particularly when the particular units called practical syllogisms by modern commentators, are set out in full. In several places Aristotle discusses them only to point out what a man may be ignorant of, when he acts faultily though well-equipped with the relevant general knowledge. It is not clear from his text whether he thinks a premise must be before the mind (' contemplated ') in order to be ' used ', nor is it of much interest to settle whether he thinks so or not. Generally speaking, it would be very rare for a person to go through all the steps of a piece of practical reasoning as set out in conformity with Aristotle's models, saying e.g. ' I am human ', and ' Lying on a bed is a good way of resting '. This does occur sometimes, in cases like his ' dry foods ' example: think of a pregnant woman deciding to eat some vitaminous food. But if Aristotle's account were supposed to describe actual mental processes, it would in general be quite absurd. The interest of the account is that it describes an order which is there whenever actions are done with intentions; the same order as I arrived at in discussing what 'the intentional action' was, when the man was pumping water. I did not realise the identity until I had reached my results; for the starting points for my enquiry were different from Aristotle's, as is natural for someone writing in a different time. In a way, my own construction is as artificial as Aristotle's; for a series of questions 'Why?' such as I described, with the appropriate answers, cannot occur very often.

43. Consider a question 'What is the stove doing?', with the answer 'Burning well' and a question 'What is Smith doing?' with the answer 'Resting'. Would not a *parallel* answer about Smith really be 'breathing steadily' or perhaps 'lying extended on a bed'? Someone who was struck by this might think it remarkable that the same expression 'What is—doing?' should be understood in such different ways: here is a case of the 'enormously complicated tacit conventions' that accompany our understanding of ordinary language, as Wittgenstein said in the *Tractatus*. And 'resting' is pretty close to lying on a bed; such a description as 'paying his gas bill', when all he is doing is handing two bits of paper to a girl, might make an enquirer say: 'Description of a human action is something enormously complicated, if one were to say what is really involved in it—and yet a child can give such a report!' And similarly for 'preparing a massacre', which would be a description of what our Nazi was doing when he was dragging metal objects about or taking ammunition out of a drawer. Aristotle's 'practical reasoning' or my order of questions 'Why?' can be looked at as a device which reveals the order that there is in this chaos.

44. Let us now consider someone saying 'If I do *this*, this will happen, if *that*, this other thing; so I'll do this'. There are three cases to consider.

(*a*) The man has no end in view. E.g. let him be considering two different foods; one is rich in vitamins, the other rich in protein; both are therefore good (i.e. wholesome). But he has no practical premise: ' Vitaminous and protein-rich foods are good for a man ': he just eats what he wants to without considering such matters. Now someone says: ' If you have some of this dish, you will get vitamins, if of that, you'll get protein ' and he says: 'All right, I'll have some of the first one '. Asked why he chose that, he might say ' Oh, I thought I'd get some protein in me '. Now this is not a case of ' practical reasoning '. If, thinking ' if I do this, this will happen ' he decides to do it, and *so* determines ' this ' as the result he wants, which before was undetermined, and if ' this' is not wanted with a view to any further end, he is not ' reasoning with a view to an end ' at all. He *could* simply not trouble to eat anything, or eat some highly unsuitable food instead, without abandoning any end. And the explanation ' Oh, I just thought I'd have something full of vitamins ' or ' Oh, I thought I'd eat some thoroughly unsuitable food ' is an extended form of what we are already acquainted with: ' I just thought I would '.

(*b*) A man who has an end in view, e.g. to eat only wholesome food, is always confronted with only one wholesome dish, and recognizing it as a kind of food that is wholesome, he takes it and not any other.

(*c*) The same man has a choice of different kinds of wholesome dishes whenever he wants to eat, and chooses some of them, but never takes others. Now *which* he chooses is not determined by his end; but he is not in the position of the first man; although he is now determining which he wants (protein or vitamin let us say), which was not predetermined, still he must choose among them or give up his objective of eating only wholesome food.

This trivial case (*c*) is an example of what is by far the most common situation for anyone pursuing an objective. Let someone be building a house, for example; his plan may not determine whether he has sash or casement windows; but he must decide which kind of window to have, at least when he comes to it, or the house will not get finished. And his calculation ' if I choose this, this will be the result, if that, that; so I'll have this ' is calculation with a view to an end—namely, the completed house; even though both alternatives would have fitted his plan. He is choosing *an* alternative that fits, even though it is not the only one that would.

G

45.   We can now consider ' practical knowledge '. Imagine someone directing a project, like the erection of a building which he cannot see and does not get reports on, purely by giving orders.   His imagination (evidently a superhuman one) takes the place of the perception that would ordinarily be employed by the director of such a project.   He is not like a man merely considering speculatively how a thing might be done; such a man can leave many points unsettled, but this man must settle everything in *a* right order. *His* knowledge of what is done is practical knowledge.

But what is this ' knowledge of what is done '?   First and foremost, he can say what the house is like.   But it may be objected that he can only say ' This is what the house is like, if my orders have been obeyed '.   But isn't he then like someone saying ' This—namely, what my imagination suggests—is what is the case if what I have imagined is true '?

I wrote ' I am a fool ' on the blackboard with my eyes shut. Now when I said what I wrote, ought I to have said: this is what I am writing, if my intention is getting executed ; instead of simply: this is what I am writing?

Orders, however, can be disobeyed, and intentions fail to get executed.   That intention for example would not have been executed if something had gone wrong with the chalk or the surface, so that the words did not appear.   And my knowledge would have been the same even if this had happened.   If then my knowledge is independent of what actually happens, how can it be knowledge of what does happen?   Someone might say that it was a funny sort of knowledge that was still knowledge even though what it was knowledge of was not the case!   On the other hand Theophrastus' remark holds good: ' the mistake is in the performance, not in the judgment '.

Hence we can understand the temptation to make the real object of willing just an idea, like William James.   For that certainly comes into being; or if it does not, then there was no willing and so no problem.   But we can in fact produce a case where someone effects something just by saying it is so, thus fufilling the ideal for an act of will as perfectly as possible.   This happens if someone admires a possession of mine and I say ' It's yours!', thereby giving it him. But of course this is possible only because property is conventional.

46. But who says that what is going on is the building of a house, or writing 'I am a fool' on the blackboard? We all do, of course, but why do we? We notice many changes and movements in the world without giving any comparable account of them. The tree waves in the wind; the movements of its leaves are just as minute as the movement of my hand when I write on a blackboard, but we have no description of a picked-out set of movements or a picked-out appearance of the tree remotely resembling 'She wrote "I am a fool" on the blackboard'.

Of course we have a special interest in human actions: but *what* is it that we have a special interest in here? It is not that we have a special interest in the movement of these molecules—namely, the ones in a human being; or even in the movements of certain bodies—namely human ones. The description of what we are interested in is a type of description that would not exist if our question 'Why?' did not. It is not that certain things, namely the movements of humans, are for some undiscovered reason subject to the question 'Why?' So too, it is not just that certain appearances of chalk on blackboard are subject to the question 'What does it say?' It is of a word or sentence that we ask 'What does it say?'; and the description of something as a word or a sentence at all could not occur prior to the fact that words or sentences have meaning. So the description of something as a human action could not occur prior to the existence of the question 'Why?', simply as a kind of utterance by which we were *then* obscurely prompted to address the question. This was why I did not attempt in §19 to say *why* certain things should be subject to this question.

Why do we say that the movement of the pump handle up and down is part of a process whereby those people cease to move about? It is part of a causal chain which ends with that household's getting poisoned. But then so is some turn of a wheel of a train by which one of the inhabitants travelled to the house. Why has the movement of the pump handle a more important position than a turn of that wheel? It is because it plays a part in the way a certain poisonous substance gets into human organisms, and that a poisonous substance gets into human organisms is the form of description of what happens which here interests us; and only because *it* interests us would we even consider reflecting on the role of the wheel's turn in carrying the man to his fate. After all, there must be an infinity of other crossroads besides the death of these people. As Wittgenstein says ' Concepts lead us to make investigations, are the expression of our interest, and direct our interest' (*Philosophical Investigations* § 570).

So the description of something that goes on in the world as ' building a house ' or ' writing a sentence on a blackboard ' is a description employing concepts of human action. Even if writing appeared on a wall as at Belshazzar's feast, or a house rose up not made by men, they would be identified as writing or a house because of their visible likeness to what we produce— writing and houses.

47. Thus there are many descriptions of happenings which are directly dependent on our possessing the *form* of description of intentional actions. It is easy not to notice this, because it is perfectly possible for some of these descriptions to be of what is done unintentionally. For example ' offending someone '; one can do this unintentionally, but there would be no such thing if it were never the description of an intentional action. And ' putting up an advertisement upside down ', which would perhaps mostly be unintentional, is a description referring to advertisements, which are essentially intentional; again, the kind of action done in ' putting up' is intentional if not somnambulistic. Or ' going into reverse ', which can be intentional or unintentional, is not a concept that would exist apart from the existence of engines, the description of which brings in intentions. If one simply attends to the fact that many actions can be either intentional or unintentional, it can be quite natural to think that events which are characterisable as intentional or unintentional are a certain natural class,' intentional ' being an extra property which a philosopher must try to describe.

In fact the term ' intentional ' has reference to a *form* of description of events. What is essential to this form is displayed by the results of our enquiries into the question ' Why? ' Events are typically described in this form when 'in order to' or 'because' (in one sense) is attached to their descriptions: 'I slid on the ice because I felt cheerful'. 'Sliding on ice' is not itself a type of description, like 'offending someone', which is directly dependent on our possessing the form of description of intentional actions. Thus we can speak of the form of description 'intentional actions', and of the descriptions which can occur *in* this form, and note that of these some are and some are not dependent on the existence of this form for their own sense.

The class of such descriptions which *are* so dependent is a very large, and the most important, section of those descriptions of things effected by the movements of human beings which go to make up the history of a human being's day or life. A short list of examples of such descriptions should bring this out. I assume a whole body as subject, and divide the list into two columns; the left hand one contains descriptions in which a happening may be intentional or unintentional, the right hand one those which can only be voluntary or intentional (except that the first few members could be somnambulistic).

Intruding	Telephoning
Offending	Calling
Coming to possess	Groping
Kicking (and other descriptions	Crouching
connoting characteristically	Greeting
animal movement)	Signing, signalling
Abandoning, leaving alone	Paying, selling, buying
Dropping (transitive),	Hiring, dismissing
holding, picking up	Sending for
Switching (on, off)	Marrying, contracting
Placing, arranging	

The role of intention in the descriptions in the right hand column will be obvious; 'Crouching' will probably be the only one that occasions any doubt. The left hand column will strike anyone as a very mixed set. Both include things that can, and things that cannot, be done by animals; something involving encounters with artefacts, like switching on or off, can of course be effected by an inanimate object; but the description only exists because we make switches to be switched on and off.

With what right do I include other members in this list? They are all descriptions which go beyond physics: one might call them vital descriptions. A dog's curled tail might have something stuck in it, but that of itself would not make us speak of the dog as holding the object with its tail; but if he has taken between his teeth and kept there some moderate-sized object, he is holding it. To speak of the wind as picking things up and putting them down again is to animalize it in our language, and so also if we speak of a cleft in rocks as holding something; though not if we speak of something as held there by the cleft. Trees, we may say, drop their leaves or their fruit (as cows drop calves); this is because they are living organisms (we should never speak of a tap as dropping its drips of water), but means no more to us than that the leaves or fruit drop off them. These descriptions are all basically at least animal. The ' characteristically animal movements ' are movements with a normal role in the sensitive, and therefore appetitive, life of animals. The other descriptions suggest backgrounds in which characteristic things are done—e.g. the reactions to an intruder.

Since I have defined intentional action in terms of language —the special question ' Why? '—it may seem surprising that I should introduce intention-dependent concepts with special reference to their application to animals, which have no language. Still, we certainly ascribe intention to animals. The reason is precisely that we describe what they do in a manner perfectly characteristic of the use of intention concepts: we describe what *further* they are doing *in* doing something (the latter description being *more* immediate, nearer to the merely physical): the cat is stalking a bird *in* crouching and slinking along with its eye fixed on the bird and its whiskers twitching. The enlarged description of what the cat is doing is not all that characterises it as an intention (for enlarged descriptions are possible of any event that has describable effects), but to this is added the cat's perception of the bird, and what it does if it catches it. The two features, knowledge and enlarged description, are quite characteristic of description of intention in acting. Just as we naturally say ' The cat thinks there is a mouse coming ', so we also naturally ask: Why is the cat crouching and slinking like that? and give the answer: It's stalking that bird; see, its eye is fixed on it. We do this, though the cat can utter no thoughts, and cannot give expression to any knowledge of its own action, or to any intentions either.

48. We can now see that a great many of our descriptions of events effected by human beings are *formally* descriptions of executed intentions. That this is so for descriptions of the type in the right hand column is evident enough. But this might be explained by saying that intention is required (as an extra feature) by the definitions of the concepts employed. This, it might be said, is no more than a quasi-legal point, or even an actual one in the case of marriage, for example. But even here it might strike someone as curious that in general special proof of intention is not required; it is special proof of lack of it (because one of the parties did not know the nature of the ceremony, for example) that would invalidate a marriage.

Surprising as it may seem, the failure to execute intentions is necessarily the rare exception. This seems surprising because the failure to achieve what one would finally like to achieve is common; and in particular the attainment of something falling under the desirability characterisation in the first premise. It often happens for people to do things for pleasure and perhaps get none or little, or for health without success, or for virtue or freedom with complete failure; and these failures interest us. What is necessarily the rare exception is for a man's performance in its more immediate descriptions not to be what he supposes. Further, it is the agent's knowledge of what he is doing that gives the descriptions under which what is going on is the execution of an intention.

If we put these considerations together, we can say that where (*a*) the description of an event is of a type to be formally the description of an executed intention (*b*) the event is actually the execution of an intention (by our criteria) then the account given by Aquinas[1] of the nature of practical knowledge holds: Practical knowledge is ' the cause of what it understands ', unlike ' speculative ' knowledge, which ' is derived from the objects known '. This means more than that practical knowledge is observed to be a necessary condition of the production of various

[1] *Summa Theologica*, Ia IIae, Q3, art. 5, obj. 1.

results; or that an idea of doing such-and-such in such-and-such ways is such a condition. It means that without it what happens does not come under the description—execution of intentions—whose characteristics we have been investigating. This can seem a mere *extra* feature of events whose description would otherwise be the same, only if we concentrate on small sections of action and slips which can occur in them.

'Practical knowledge' is of course a common term of ordinary language, no doubt by inheritance from the Aristotelian philosophy. For that philosophy has conferred more terms on ordinary language than any other, in senses more, or less, approximating to those of Aristotle himself: 'matter', 'substance', 'principle', 'essence' come readily to mind; and 'practical knowledge' is one of them. A man has practical knowledge who knows how to do things; but that is an insufficient description, for he *might* be said to know how to do things if he could give a lecture on it, though he was helpless when confronted with the task of doing them. When we ordinarily speak of practical knowledge we have in mind a certain sort of general capacity in a particular field; but if we hear of a capacity, it is reasonable to ask what constitutes an exercise of it. E.g., if my knowledge of the alphabet by rote is a capacity, this capacity is exercised when I repeat these noises, starting at any letter. In the case of practical knowledge the exercise of the capacity is nothing but the doing or supervising of the operations of which a man has practical knowledge; but this not *just* the coming about of certain effects, like my recitation of the alphabet or of bits of it, for what he effects is formally characterised as subject to our question 'Why?' whose application displays the A—D order which we discovered.

Naturally my imaginary case, in which a man directs operations which he does not see and of which he gets no information, is a very improbable one. Normally someone doing or directing anything makes use of his senses, or of reports given him, the whole time: he will not go on to the next order, for example, until he knows that the preceding one has been executed, or, if he is the operator, his senses inform him of what is going on. This knowledge is of course always 'speculative' as opposed to 'practical'. Thus in any operation we really can speak of two knowledges—the account that one could give of what one was doing, without adverting to observation; and the account of exactly what is happening at a given moment (say) to the material one is working on. The one is practical, the other speculative.

Although the term ' practical knowledge ' is most often used in connexion with specialised skills, there is no reason to think that this notion has application only in such contexts. ' Intentional action ' always presupposes what might be called ' knowing one's way about ' the matters described in the description under which an action can be called intentional, and this knowledge is exercised in the action and is practical knowledge.

49. The distinction between the voluntary and the intentional seems to be as follows: (1) Mere physical movements, to whose description our question ' Why? ' is applicable, are called voluntary rather than intentional when (a) the answer is e.g. ' I was fiddling ', ' it was a casual movement ', or even ' I don't know why ' (b) the movements are not considered by the agent, though he can say what they are if he does consider them. It might seem that this is a process of empirical discovery; for example, a man who wanted to say what movements he made in detail might go through the motions in order to find out. Isn't the knowledge so gained observational? That it is not can be seen if we remember that he does not necessarily have e.g. to *look* at his hands in order to say; and it is even possible to make this discovery by going through the motions (e.g. of tying a knot) in imagination, but imagination could never have authority to tell us what would be the observed result of an experiment. (2) Something is voluntary though not intentional if it is the antecedently known concomitant result of one's intentional action, so that one could have prevented it if one would have given up the action; but it is not intentional: one rejects the question ' Why? ' in its connexion. From another point of view, however, such things can be called involuntary, if one regrets them very much, but feels ' compelled ' to persist in the intentional actions in spite of that. (3) Things may be voluntary which are not one's own doing at all, but which happen to one's delight, so that one consents and does not protest or take steps against them: as when someone on the bank pushes a punt out into the river so

that one is carried out, and one is pleased.—'Why' it might be asked, 'did you go sliding down the hill into that party of people?' to which the answer might be 'I was pushed so that I went sliding down the bank'. But a rejoinder might be 'You didn't mind; you didn't shout, or try to roll aside, did you?' (4) Every intentional action is also voluntary, though again, as at (2), intentional actions can also be described as involuntary from another point of view, as when one regrets 'having' to do them. But 'reluctant' would be the more commonly used word.

50. I have completed the enquiry into intentional action and intention with which an action is done, and will now return to the topic I left at §4: expression of intention for the future. What I have said about intention in acting applies also to intention in a proposed action. And, indeed, quite generally, the applicability of the question 'Why?' to a prediction is what marks it out as an expression of intention rather than an estimate of the future or a pure prophecy. But what distinguishes it from a hope? A hope is possible even concerning one's own future intentional actions: 'I shall be polite to him—I hope'. Grounds of hope are mixed of reasons for wanting, and reasons for believing that the thing wanted may happen; but grounds of intention are only reasons for acting.

51. A possible answer to the question 'Why?' about an expression of intention regarding a future action is 'I just want to, that's all'. This form of words is of course possible in relation to a present action too. But its significance appears to change according as it is said of a present, or of a future, action. Said of a present action, it suggests an objection to being troubled with questions: this is just what I am doing, and I am not interested in having it queried. But this does not mean that the question 'Well, at least what's pleasant or interesting about it?' is shewn to have no application. What is the man at in doing the thing that he 'just wants to'? Whiling away the time? Seeing if he can finish some futile thing which for a moment's idle occupation he has started—as one might persist in seeing if one could find all the letters of the alphabet on a small bit of news-

paper? ' I want to ' is not an explanation of something that a man *is* doing.

It is different with a proposed action. My remarks about ' wanting ' an object or a state of affairs at §37 do not necessarily apply to wanting to do something. Say I notice a spot on the wall-paper and get out of my chair. Asked what I am doing I reply ' I'm going to see if I can reach it by standing on my toes '. Asked why, I reply ' I want to, that's all ' or ' I just had the idea '. Here I may be excluding the idea that there is any further point, any room for more answers from me; and no one can say: But there is a place for an answer of a certain type, which place requires to be filled. But if I stay there with my finger on the spot, or keep on reaching up to it, and when asked why, I say ' I want to, that's all ', there does seem to be a gap demanding to be filled. What am I doing? Am I e.g. seeing how long I can keep it up? It is not just a matter of eccentricity. The question is, what information ' I want to do it, that's all ' gives you, apart from the fact that I am doing it: what it tells you that ' No particular reason ' would not tell you. For it is certainly not a report that a feeling of desire is animating me in connexion with what I am doing.

But if an idea of something I might do inspires me to set out to do it, or to make up my mind to do it, not with any end in view, and not as anything but itself, this is ' just wanting ' to do it; and to say ' I just want to, that's all ' is to explain that that is the situation.

' I wanted to, that's all ' might tell us that had *had* been the situation when I did something. And one can say ' I wanted to ' of a present action.

We could imagine a special mood of verbs (compare the ' optative ' mood in Greek) in which the future tense was used purely to express intention of doing something just because one wants to, and a ' past future ', as it were, in the same mood used in place of ' I wanted to '. But there would be no present of this mood, if this were its function.

This ' I want, that's all ' applies only to doing.

52. Let us consider ' I am going to do it ' said as an expression of intention, and ' I am not going to do it ' as a belief on

evidence—when the ' it ' is one and the same.

' I am going for a walk—but shall not go for a walk ' is a contradiction of a sort, even though the first part of the sentence is an expression of intention, and the second an estimate of what is going to happen. Suppose there are no difficulties about the man's going for a walk? How can he say both things, and claim that there is no contradiction because one part is just an expression of intention and the other judgment on what will actually happen?

The contradiction consists in the fact that if the man does go for a walk, the first prediction is verified and the second falsified, and vice versa if he does not go. And yet we feel that this is not, so to speak, a head-on contradiction, like that of pairs of contradictory orders, contradictory hypotheses, or opposed intentions.

If I say I am going for a walk, someone else may know that this is not going to happen. It would be absurd to say that *what* he knew was not going to happen was not the very same thing that I was saying *was* going to happen.

Nor can we say: But in an expression of intention one isn't saying anything is going to happen! Otherwise, when I had said ' I'm just going to get up ', it would be unreasonable later to ask ' Why didn't you get up? ' I could reply: ' I wasn't talking about a future happening, so why do you mention such irrelevancies? '

Ought one really always to say ' I am going to . . . unless I am prevented '? or at least to say that there is an implicit ' unless I am prevented ' (an implicit *deo volente*) in every expression of intention? But ' unless I am prevented ' does not normally mean ' unless I do not do it '. Suppose someone said ' I am going to . . . unless I am prevented, or I change my mind '?

In the small activities of everyday life, to say ' I am going to, unless I am prevented ' would be absurd, like putting ' unless my memory deceives me ' after every report one gave of what had happened. And yet there are cases in which one's memory deceives one. One may therefore think: in those cases it would have been more correct for one to add ' unless my memory deceives me ' to the report. But there is no way of choosing the right cases; for one would actually choose them when for particular reasons there was some doubt about the report; well, we can suppose that a man never makes a confident report when he has any special reason to doubt, but this man will probably still sometimes be wrong in what he confidently reports. We know this because we all are sometimes wrong. But this general ground could only lead one to add ' unless my memory deceives me ' to every report. It would then be no more than an acknowledgement that ' in every case, one *could* be wrong '— which does not mean ' one could be wrong in every case '. When one considers a particular case—e.g. ' I met so-and-so yesterday ' —one is inclined to say ' I *couldn't* be wrong '. But even if one made a habit of asking ' Can I say ' I *couldn't* be wrong ' in that way? ' before venturing on a report, one would probably have to concede later that sometimes one had been wrong; at least one could not say that this possibility is ruled out for anyone who adopts this habit, for people sometimes are wrong about what they are quite certain of. So that all one is really saying is: in this case I *am* not wrong—i.e.: it happened. And one is sometimes wrong, but mostly right.

Similarly, when one says ' I am going to ' one may always be prevented but need not consider that; mostly, one is not prevented. And it would be useless to try to attach ' unless I am prevented ' to the right cases, in which one actually is prevented but there was no reason to expect it. In saying ' I am going to ', one really is saying that such-and-such is going to happen . . . which may not be true.

But if one is considering the fact that one may not do what one is determined to do, then the right thing to say really *is* ' I am going to do this . . . unless I do not do it '. Even ' I am going (or not going) to do this, unless I am prevented, or change my mind ' is not adequate, as can be seen from the case of St. Peter, who did not change his mind about denying Christ, and was not prevented from carrying out his resolution not to, and yet did deny him.

' I am going to . . . unless I do not ' is not like ' This is the case, unless it isn't'. It has an analogue in estimates of the future: ' This is going to happen . . . unless it doesn't '. (Someone may prevent it.) This could be said even of an eclipse of the sun; because the verification of predictions awaits the event—and the sun might blow up before the eclipse.

It is for this reason that in some cases one can be as certain as possible that one will do something, and yet intend not to do it. So a man hanging by his fingers from a precipice may be as certain as possible that he must let go and fall, and yet determined not to let go. Here, however, we might say: ' In the end his fingers let go, not he '. But a man could be as certain as possible that he will break down under torture, and yet determined not to break down. And St. Peter might perhaps have calculated 'Since *he* says it, it is true'; and yet said 'I will not do it '. The possibility in *this* case arises from ignorance as to the way in which the prophecy would be fulfilled; thus St. Peter could do what he intended not to, without changing his mind, and yet do it intentionally.